Number System in Saṁskṛit

*

Dr. Ramamurthy N

M.Sc., B.G.L., C.A.I.I.B., C.C.P., D.S.A.D.P., C.I.S.A., P.M.P., CGBL, Ph.D.

*

Published by

CBH PUBLICATIONS

*

Name of the Book: **Number System in Saṁskṛit**

First Edition: 2014

Author: **Dr. Ramamurthy N.**
http://ramamurthy.jaagruti.co.in

ISBN (13): 978-93-82237-04-4

Price: **₹. 190/-**

Published by: **CBH PUBLICATIONS**
51/4 Sriganesh, North Aman Coil Street,
Kamarajar Nagar,
Veturnimadom P.O.,
Nagercoil 629 003
phone: 04652-222813
e-mail: cbhpub@gmail.com
http://cbhpublications.co.in/

*****

TABLE OF CONTENTS

*****

Dedication

Mātru Devo Bhava

I dedicate this book with *pranāms* to my mother *Śreemati* **Alankāravalli Ammal**, who brought me to this level right from scratch. She is the Divine Mother *Śree Devee* in the human form in this world. Though she is not physically available to see me at this glassy, I am sure all these are possible only with her blessings and she continues to live with me spiritually.

Dr. Ramamurthy N

Prof. S.A.SANKARANARAYANAN, Retd
Tamilnadu Educational Service
Consultant Comparatist,Scripture Translation;
Sri Sankara Arts and Science College, Kamakoti Trust,
(Affiliated to Bharathidasan University;

Truechitrambalam

You gave me you and took me instead,
O! Sankara, who of us is squarely smart?
Bliss unending I got thereof;
What have you got from me in turn?
O, My Lord, en-templed in the conscious!
O, Civa,Abiding at holy Perunthurai!
O, My Father, Supreme parked in my flesh!
None am I meet for such an exchange!
Tiruvaachakam-Koyil Tiruppathikam

Editor Emeritus, The Scholar Miscellanist;
Executive Editor, Oriental Journal of Asian Studies,
SASTRA University, TN;

Formerly Prof and Head, English,Srinivasa Ramanujan Centre,
SASTRA, Kumbakonam.

Resource personnel BDU, UGC Academic Staff College,
Directorate of Translation Studies

Date: 18th July 2013

Foreword

"Language must embody what it indicates"-
Alfred North Whitehead

To text much *'mathema'* in a little, *multum in parvo*, sleep is implicated: by derivation, *jāgrat* is adverted. With a primer of an exclusive kind to which this work belongs, from the post-modern age of hypnoglyph, may the innocents in many of us wake up into a realm of a larger Time vectored *'Vedic'*, with modalities of dormancy impounded in it by the superstructures of the Western, compliant or not!

If all knowledge be acknowledgement, as Plato would have it, it is good to be ekphrastic like Mr. **N.**

Res:73/74,Mettu Street,Kumbakonam,
PIN 612001 TN, INDIA.
Ph:0435-2424819; Mob.9442211501
e-mail:profsas2003@gmail.com

Ramamurthy, by indirection, often tabling that number is neither revelry nor devilry, or geometry is none of the debauchery or carousal of Space. Mere mention of *Satapatabrāhmana* and such and such is gnomic to suggest that when our sleep is profound, we stand and are put in hold! *Hu'pnos hedraios, orthoi, nustagmo's* [Epidemics 6.4.15; Littre':310].

My purport is to declaim, though it might seem a bit bizarre, by an analogy: to use the surmise of Galen (129-199 C.E), the late Classical successor of Hippocrates, and the so called advanced most wakeful number-theoreticians of our day may achieve a soothing slumber if they remain erect and vigilant until they begin to nod and wink!

A saying goes: Even Homer nods! If we return to origins, in peripety, we experience a jolt or an un-hinged-ness of our advanced-ness: a constraint to sleepwalk, awakening only when stumble on a stone, paleolithic! I mean, tit-bits are misnomers for even if they pass for so so, cover they do an aura of a decisive juncture of an essential moment, the author attempts to recreate, tending to be minimally narrational. Of course, Laplace supports the choice of Mr. Ramamurthy's locution *'pragnantesten'*, most pregnant, for a total emergence of *mathema*'s progeny from parent Vedic, with a historic-Indic sense and demystification of not just instants but incidents, relating ensembles of Time and Space. The tit-bit type of primering the profound is endemic to Mr. Ramamurthy's, for his is spiring fulfilled continuity of *Shree Chakra* resolution.

The contents of this book are a phalanx in an under-coding for a spirited adoption of a subtle differential from our era's norms. Come what may, if a lay starter is schooled in these hundred and odd pages of ages of labor, the durable circadian boundaries blur. Mathematical logic and mathematical rhetoric as more than mere approval of a diaristic discourse must help the learner thank the impersonal chronicle of the theological track of a believer's subjectivity, engendering no muddle in the middle we owe to!

Ramamurthy in his more-than-a-text-book seems to postulate a categorical neither-here-nor-there-ness of what the student world of mathematics has been in. Any way the book is not to be viewed as a recordation calendrically, though there is an overtone of temporal character in the signification of anamnesis proper to the syllabus of the concerned discipline.

As one devoted to Oriental Studies, the polymathy of the author negotiates the hermeneutic quandary by the surprising inclusion of *Lalitā Sahasranāma*. Salutary this is; for mathema as a system of number in *Samskrit* is at once existence and existent. A certain anonymous vigilance leads the initiate into a tan inverse relation of two adjacencies. Salute we may the restorative energy of this vigilance which approves of our withdrawal from the corybantic refuge from the naive arithmetic world supposedly! Not that I present, nor rather compliment, a *de facto* eulogy of what we have been sleeping over. The moot question hovers: Is the system as such a commitment in the uncommittedness of what has long been in wait to get acknowledged?

Ramamurthy's *vade mecum*, yes, I would call it so, on *Vedic* et. al. *numina* move from the theoretical vantage point: a harboring stratum of core parody. No longer could it be: 'Calculus, thy name is Newton if not Leibnitz'; for the rudiments in the oversight of occidental inculcation, had never been steered through.

I understand, as would perhaps Spengler or Schrodinger do, that Ramamurthy's effort has, for once, disclosed the infiltration of a kinetic imperative into a bygone facet of impermeable existence, almost like an un-diapered premise of representation entrenching appropriation by Western historicism.

Reading this primer shall tend to generate humility in readership without an accompanying pride of mollification or closure. I wonder and exclaim from some safe distance: "and after so many centuries, we don't know much about this subject". Though it sounds anticlimactic, prefer I do, to put in black and white, as if from a third world, that all advancedness are an accumulation, sadly incompatible with the laden intimations aesthetic, philosophical and religious already there, only to be reminisced by a chosen few of whom Mr. Ramamurthy is for sure, the one.

I commend this Primer Text as one of supreme importance to be a reference amidst an enmeshing dormancy our syllabus is seldom free from and am glad to meditate on its admission to the Jewish Congress Library Catalog by virtue of its exuberance.

"The head sublime, the heart pathos, the genitals beauty, the hands and feet proportion.

As the air to a bird or the sea to a fish,

So is contempt to the contemptible.

The crow wished everything was black,

the owl that everything was white.

Exuberance is Beauty" . . .

-William Blake

OM TAT SAT

Prof. *S.A.Sankaranarayanan*

*****

Introduction

It is beyond doubt that Mathematics, what we have today, owes a huge debt to the outstanding contributions made by Indian Mathematicians over many hundreds, probably thousands, of years. What is quite surprising is that there has been a reluctance to recognise this and one has to conclude that many famous historians of Mathematics found, what they expected to find, or perhaps even what they hoped to find, rather than to realise what was so clear in front of them. The Indian Mathematics was not found just by calculations; Indian Mathematicians could codify it through penance and deep wisdom. The latest evidence in this regard is Vedic Mathematics of 20th Century C.E., obtained through penance by HH Jagadguru Swāmi Śrī Bhārati Kṛṣṇa Tīrthāji Mahāraj. We can boldly and loudly claim that all concepts of modern day Mathematics have analogues in Indian Mathematics.

The "huge debt" is the beautiful number system invented by the Indians on which most of the Mathematical developments have rested. Laplace puts this as:

“The ingenious method of expressing every possible number using a set of ten symbols (each symbol having a place value and an absolute value) emerged in India. The idea seems so simple nowadays that its significance and profound importance is no longer appreciated. Its simplicity lies in the way; it facilitated calculation and placed arithmetic foremost amongst useful inventions. The importance of this invention is more readily appreciated, when one considers that it was beyond the two greatest men of antiquity – Archimedes and Apollonius”.

Histories of Indian Mathematics used to begin by describing the Geometry contained in the *Sulbasootras* but research into the history of Indian Mathematics has shown that the essentials of this

Geometry were older being contained in the altar constructions described in the *Vedic* mythology text – *Shatapatha Brāhmana* and the *Taittreeya Samhita.* Also it has been shown that the study of Mathematical Astronomy in India goes back to at least the third millennium B.C.E. and Mathematics and Geometry must have existed to support this study in these ancient times.

The first Mathematics was developed in the Indus valley. The earliest known urban Indian culture was first identified at Harappa in the Punjab and then at Mohenjo-Daro, near the Indus River in the Sindh. Both these sites are now in Pakistan but this is still covered by the term "Indian Mathematics" which, refers to Mathematics developed in the Indian sub-continent. The Indus civilisation (or Harappa civilisation as it is sometimes called) was based in these two cities and also in over a hundred small towns and villages around. It was a civilisation, which began around 2500 B.C.E., and survived until 1700 B.C.E., or even later. The people were literate and used a written script containing around 500 characters, which some have claimed to have deciphered but, being far from claim that this is the case, much research remains to be done before a full appreciation of the Mathematical achievements of this ancient civilisation can be fully assessed.

It is so sad that the due credit has not been provided to Indian Mathematicians and to add fuel to the same, some of the IP or copyrights or fame has been provided to later people who gave the Mathematical concepts, which are already available in Indian Mathematics. One important reason is that most of these texts of Indian Mathematics were given in cryptic form and to decipher the same the students need a deep knowledge of Saṁskṛit and still further knowledge in Mathematics.

At the outset when the author got some feedback on this book as it talks on various things at a very high level like tit-bits. No single volume can be written comprehensively covering entire Indian Mathematics. The very aim of this book is to provide a high level idea of Indian Mathematics and to demystify some of the Mathematical representations and explain the same in simple terms. In this process an attempt has been made to give an overall history of Indian Mathematics also. Again an attempt has been made to, at least, list down the Luminaries of Indian Mathematics in an annexure.

Any discussion involving Mathematics and Saṁskṛit is incomplete without mentioning about Vedic Mathematics. Hence an introductory chapter on Vedic Mathematics has also been included in this book.

Conventions used in this book: The transliterated Saṁskṛit words are written in italics – for instance *Bhoota*. When Saṁskṛit words are transliterated in English diacritical marks are normally used. However the same has not been used in its entirety in this book, since it makes the reading a little more difficult and since this book is intended for common audience.

Many many thanks from the bottom of my heart are due to Prof. S.A.Sankaranarayanan, who has written a nice introduction and some pleasantries about the author and this edition. I will be failing in my duty if I do not thank *Shree*. M. Easwaran, who has printed and published this book. He not only does the publication tasks, but also reviews the book and gives vital feedback so that the book is in the fashion it is at present. Even amidst all constraints like soaring power cut, etc., nice job done – thanks a lot. Sincere thanks are due to all those who supported in this noble cause.

Om Tat Sat

Chennai
November 2013

Dr. *Ramamurthy N*

1. History of Mathematics in India

"The *Vedas* were the beginning of our spiritual knowledge; the *Vedas* will remain till the end. These compositions of an unknown antiquity are as many breasts of the eternal Mother of knowledge from which our succeeding ages have all been fed". *Śree* Aurobindo

Egyptians and Babylonians were thought of being the height of civilisation and of Mathematical skills around the period of the Indus civilisation, yet V G Childe in "New Light on the Most Ancient East" (1952) writes as:

"India confronts Egypt and Babylonia by the 3rd millennium with a thoroughly individual and independent civilisation of her own, technically the peer of the rest. And plainly it is deeply rooted in Indian soil. The Indus civilisation represents a very perfect adjustment of human life to a specific environment. And it has endured; it is already specifically Indian and forms the basis of modern Indian culture".

It is very clear that Harappans had adopted a uniform system of weights and measures. Several scales for the measurement of length were also discovered during excavations. One was a decimal scale based on a unit of measurement of 1.32 inches (3.35 centimetres) which has been called the "Indus inch". Of course ten units are then 13.2 inches, which is quite believable as the measure of a 'foot'. A similar measure based on the length of a foot is present in other parts of Asia and beyond. Another scale was discovered when a bronze rod was found which was marked in lengths of 0.367 inches. The accuracy with which these scales are marked is certainly surprising. Now 100 units of this measure is 36.7 inches, which is the measure of a stride. Measurements of the ruins of the buildings, which have been excavated show that these units of length were accurately used by the Harappans in construction.

It is unclear exactly what caused the decline in the Harappan civilisation. Historians have suggested four possible causes:

- A change in climatic patterns and a consequent agricultural crisis;
- A climatic disaster such as floods or severe drought, disease spread by epidemic, etc.
- Invasion of Indo-Aryan people from the North.
- The favourite theory used to be the last of the four, but recent opinions favour one of the first three.

What is certainly true is that eventually the Indo-Aryan people from the North did spread over the region. This brings the earliest literary record of Indian culture, the *Vedas*, which were composed in *Vedic* Saṁskṛit, probably between 1500 B.C.E., and 800 B.C.E. At first these texts, consisting of hymns, spells and ritual observations, were transmitted orally – hence the name *Shruti*. Later the texts became written works for use of those practicing the *Vedic* religion.

The next Mathematics of importance on the Indian subcontinent was associated with religious texts. It consisted of the *Sulbasootras*, which were appendices to the *Vedas* giving rules for constructing altars. They contained quite an amount of geometrical knowledge, but the Mathematics was a developed one, not for its own sake, but purely for practical religious purposes.

The main *Sulbasootras* were composed by Boudhāyana (about 800 B.C.E.), Mānava (about 750 B.C.E.), Āpastamba (about 600 B.C.E.) and Kātyāyana (about 200 B.C.E.). These men were both priests and scholars but they were not Mathematicians in the modern sense. Although there is no information on these men other than the texts they wrote, they have been considered as scholarly Mathematicians. There is another scholar, who again was not a Mathematician in the usual sense, who lived around this period. That was Pānini who achieved remarkable results in his studies of Saṁskṛit grammar. There could be a reasonable question what Saṁskṛit grammar has to do with Mathematics. It certainly has something to do with modern theoretical Computer Science, for a Mathematician or Computer Scientist working with formal language theory will recognise just how modern some of Pānini's ideas are[1].

Before the end of the period of the *Sulbasootras*, around the middle of the 3rd Century B.C.E., the Brāhmi numerals had begun to appear. The below is one style of the Brāhmi numerals.

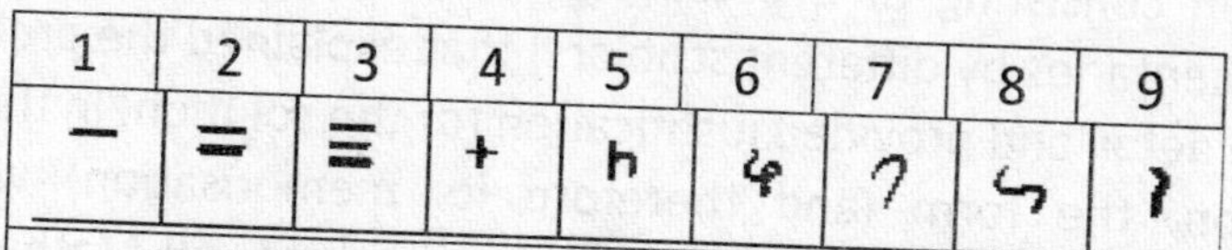

Brāhmi numerals around 1st Century C.E.

These are the earliest numerals which, after a multitude of changes, eventually developed into the numerals 1, 2, 3, 4, 5, 6, 7, 8, 9 used today. It may be noted that '0' is not yet known to the Mathematicians and there are only nine digits.

Indian Mathematics emerged in the Indian subcontinent from 1200 B.C.E., until the end of the 18th Century. In the classical period of Indian Mathematics (400 C.E. to 1200 C.E.), important contributions were made by scholars like Āryabhaṭa, Brahmagupta and Bhaskara

[1] Forbes Magazine, Volume 6 dated 05th May 2006 published an article saying that Saṁskṛit is the most convenient language for computer software programming.

II. The decimal number system in use today was first recorded in Indian Mathematics. Indian Mathematicians made early contributions to the study of the concept of zero as a number, negative numbers, arithmetic and algebra. In addition, trigonometry was further advanced in India and in particular, the modern calculation of Sine and Cosine were developed here. These Mathematical concepts were transmitted to the Middle East, China and Europe and led to further developments that now form the foundations of many areas of Mathematics.

Ancient and medieval Indian Mathematical works, all composed in Saṁskṛit, usually consisted of a section of *sootras* in which a set of rules or problems were stated with great economy in verse in order to aid memorisation by a student. This was followed by a second section consisting of a prose commentary (sometimes multiple commentaries by different scholars) that explained the problem in more detail and provided justification for the solution. In the prose section, the form (and therefore its memorisation) was not considered as important as the ideas involved. All Mathematical works were orally transmitted until approximately 500 B.C.E.; thereafter, they were transmitted both orally and in manuscript form. The oldest extant Mathematical document produced on the Indian subcontinent is the birch bark Bakhshali Manuscript, discovered in 1881 in the village of Bakhshali, near Peshawar (modern day Pakistan) and is likely from the 7th Century C.E.

A later landmark in Indian Mathematics was the development of the series expansions for trigonometric functions (Sine, Cosine and Arc Tangent) by Mathematicians of the Kerala School in the 15th Century C.E. Their remarkable work, completed two centuries before the invention of calculus in Europe, provided what is now considered the first example of a power series (apart from geometric series). However, they did not formulate a systematic theory of differentiation and integration, nor is there any direct evidence of their results being transmitted outside Kerala.

Excavations at Harappa, Mohenjo-Daro and other sites of the Indus Valley Civilisation (IVC) have uncovered evidence of the use of "practical Mathematics". The people of the IVC manufactured bricks whose dimensions were in the proportion 4:2:1, considered favourable for the stability of a brick structure. They used a standardised system of weights based on the ratios: 1/20, 1/10, 1/5, 1/2, 1, 2, 5, 10, 20, 50, 100, 200 and 500, with the unit weight equaling approximately 28 grams (and approximately equal to the English ounce or Greek uncial). They mass produced weights in regular geometrical shapes, which included hexahedra, barrels, cones and cylinders, thereby demonstrating knowledge of basic geometry.

The inhabitants of Indus civilization also tried to standardize measurement of length to a high degree of accuracy. They designed a ruler — the Mohenjo-Daro ruler — whose unit of length (approximately 1.32 inches or 3.4 centimeters) was divided into ten equal parts. Bricks manufactured in ancient Mohenjo-Daro often had dimensions that were integral multiples of this unit of length.

1.1. Mathematics and India

In India's early civilisations, the first expression of Mathematical understanding appears in the form of counting systems. Numbers in very early societies were typically represented by groups of lines, though later different numbers came to be assigned specific numeral names and symbols (as in India) or were designated by alphabetic letters (such as in Rome). Although today, we take our decimal system for granted, not all ancient civilisations based their numbers on a ten–base (decimal) system. In ancient Babylonia, a sexagesimal (base 60) system was also in vogue.

In ancient India, Mathematics or *'Gaṇita'* was the "Science of Calculations". It was primarily studied in the context of numerical computation and Geometric measurement. Most of the Indian Mathematical works can be found as a part of *Jyotiṣa* or Astronomy. This is because Astronomy – inseparable from Astrology, which dealt with the measurement of time using the heavenly bodies, involved high levels of sophisticated numerical computations. Mathematics in ancient India was so well developed that the body of knowledge was not restricted only to the elite scholars. It was prevalently used even by the common people in their daily activities and profession.

1.2. Numbers and Veda

In the *Vedas* there are descriptions of the meanings of the nine numbers. To begin with, the creation is made of nine elements and each element stands for a number. The elements are shown on a circle of nine points. The basic numbers always remain one to nine and all extra numbers are only the product of these nine stages seen at different levels:

1. The number **one** is Absolute, or God, from which everything originates. This universe, coming from the Absolute, is sustained by the Absolute and finally merges into the Absolute. This is the theory of *Advaita*. There is no two.
2. **Two** stands for unmanifested nature.
3. **Three** for the manifested nature.
4. At **four** there is the whole world of mind and the feeling of existence.

5. Number five stands for ether or space and has the quality of sound. The physical world, which we can hear, touch, see, taste and smell, starts at number **five**.
6. **Six** is air and has the quality of touch.
7. Fire is at **seven** with the quality of light.
8. Water is at **eight** with the quality of taste and bonding.
9. Finally, at **nine** is the earth. It has smell and crystalline form.
10. The number **ten** is 1 with a nought next to it: the one stands for the Absolute and the nought for the unmanifest.

The ultimate or the Absolute (*Brahmam*) is one and with the start of creation it unfolds itself into nine states and ends there.

This universe is created by the Absolute, the creation of manifested forms is sustained by it and in the end the whole creation will once again merge into it. That is the *Brahmam* and that is what one needs to know.

A story goes: Ten people were going across the country to another land and they had to cross a river. The river was shallow but the currents were swift. They managed to cross the river and after reaching the other shore, they wanted to make sure that no one was drowned. Each of them counted the others and found the total of nine only, for none of them would count self. They were sorry and disturbed. A holy man was passing along the bank and seeing them miserable he asked the reason for their worries. They narrated their plight. The holy man saw their difficulty and foolishness, so he asked all of them to line up. With his stick he hit one and separated him from the others. The next one he hit twice and then separated him. Likewise he hit the tenth man ten times and declared them ten and assured them that none was lost.

(*Śrī Ādi Śankara* – in one of his speeches – a hear-say story)

This story is told to illustrate that the tenth is **Self**. Tenth is the one with zero, which is unmanifest. One to nine are the numbers of manifestation and at ten, the same **Self** which is the one stands with unmanifest nature by its side. Further on the same, repetition of numbers occurs. The one at number ten embodies the nine manifestations within it. The creation starts with one and at ten it again stands as one with all the nine manifestations.

The above rule of 1 to 9 and the ten standing separately can also be compared to *Daśāvatāra* of Lord *Viṣnu* with 9 *Avatāras* completed and the *Kalki Avatār* yet to take place.

Just as the Absolute is inside everything, so is the number **one,** which is a factor of every number and every number is a factor of itself.

1.2.1. *Samhitās* and *Brāhmanas*

Samhitā offers prayers to various Gods/ *Devas*. *Brāhmana* explains the method of performing *yagnas*. The religious texts of the *Vedic* period provide evidence for the use of large numbers. By the time of the *Yajurvedasaṃhitā* (1200–900 B.C.E.), numbers as high as 10^{12} were being included in the texts. For instance, the *mantra* at the end of the *annahoma* performed during the *aśvamedha yagna* and uttered just before, during and just after Sunrise, invokes powers of ten upto 10^{53} with distinct names as well it talks about the fraction one-sixteenth:

Kala	1/16	One sixteenth
Kāṣṭha	1/12	One twelfth
Sapha	1/8	One eighth
Pāda	¼	Quarter or one fourth
Eka	1	One
Daṣa	10	Ten

Śata	10^2	Hundred
Sahasra	10^3	Thousand
Ayuta	10^4	Ten Thousand
Niyuta or *Lakṣa*	10^5	Hundred Thousand or One Lakh
Prayuta	10^6	Million or Ten lakhs
Adbhuta or *Koti*	10^7	Ten Million or one Crore
Nyadbhuta	10^8	Hundred Million
Samudra or *Abja*	10^9	Billion
Madhya or *Sarva*	10^{10}	Ten Billion
Nikharva	10^{11}	Hundred Billion
Mahāpadma	10^{12}	Trillion
Śanku	10^{13}	Ten Trillion
Jaladhi	10^{14}	Hundred Trillion
Antya	10^{15}	Zillion
Madhya	10^{16}	Ten Zillion
Parārdha	10^{17}	Hundred Zillion
Titilamba	10^{27}	
Sarvabala	10^{45}	
Tallakṣana	10^{53}	

The *Satapatha Brāhmana* (7th Century B.C.E.) contains rules for ritual geometric constructions that are similar to the *Sulba Sūtras*.

1.2.2. *Śulba Sūtras*

The *Śulba Sūtras* (literally means – Aphorisms of the Chords – 700–400 B.C.E.) list rules for the construction of sacrificial fire altars. Most Mathematical problems discussed in the *Śulba Sūtras* spring from "a single theological requirement" that of constructing fire altars, which have different shapes but occupy the same area. The altars were required to be constructed of five layers of burnt brick, with further condition that each layer consist of 200 bricks and that no two adjacent layers have congruent arrangements of bricks.

According to Hayashi, the *Śulba Sūtras* contain "the earliest extant verbal expression of the Pythagorean Theorem in the world, although it had already been known to the Old Babylonians". The diagonal rope (*akṣṇayā-rajju*) of an oblong rectangle produces both which the flank (*pārśvamāni*) and the horizontal (*tiryaṇmānī*) produce separately".

Since the statement is a *sūtra*, it is necessarily compressed and what the ropes produce is not elaborated on, but the context clearly implies the square areas constructed on their lengths

They contain lists of Pythagorean triples, which are particular cases of Diophantine equations. They also contain statements (of course approximately) about squaring the circle and circling the square.

Boudhāyana (8th Century B.C.E.) composed the *Boudhāyana Sulba Sūtra*, the best-known *Sulba Sutra*, which contains examples of simple Pythagorean triples, such as: (3, 4, 5), (5, 12, 13), (8, 15, 17), (7, 24, 25) and (12, 35, 37) and as well as a statement of the Pythagorean theorem for the sides of a square: "The rope which is stretched across the diagonal of a square produces an area double the size of the original square".

It also contains the general statement of the Pythagorean Theorem (for the sides of a rectangle): "The rope stretched along the length of the diagonal of a rectangle makes an area which the vertical and horizontal sides make together". Boudhāyana gives a formula for the value of $\sqrt{2}$ as:

$$\sqrt{2} = 1 + 1/3 + 1/(3 \times 4) + 1/(3 \times 4 \times 34) \approx 1.4142156$$

The formula is accurate up to five decimal places, the true value being 1.41421356. This formula is similar in structure to the formula found on a Mesopotamian tablet from the Old Babylonian period (1900–1600 B.C.E.):

$$\sqrt{2} = 1 + 24/60 + 51/60^2 + 10/60^3 = 1.41421297$$

This expresses $\sqrt{2}$ in the sexagesimal system and that too is accurate up to 5 decimal places (after rounding).

According to Mathematician S. G. Dani, the Babylonian cuneiform tablet Plimpton written in 1850 B.C.E. "contains fifteen Pythagorean triples with quite large entries, including (13500, 12709, 18541) which is a primitive triple, indicating, in particular, that there was sophisticated understanding on the topic" in Mesopotamia in 1850 B.C.E. Since these tablets predate the *Sulbasūtras* period by several Centuries, taking into account the contextual appearance of some of the triples, it is reasonable to expect that similar understanding would have been there in India".

Dani continues to say: "As the main objective of the *Sulvasūtras* was to describe the constructions of altars and the geometric principles involved in them, the subject of Pythagorean triples, even if it had been well understood may still not have featured in the *Sulvasūtras*. The occurrence of the triples in the *Sulvasūtras* is comparable to Mathematics that one may encounter in an introductory book on architecture or another similar applied area and would not correspond directly to the overall knowledge on the topic at that time. Since, unfortunately, no other contemporaneous sources have been found it may never be possible to settle this issue satisfactorily".

In all, three *Sulba Sūtras* were composed. The remaining two, the *Manava Sulba Sūtra* composed by Manava (750–650 B.C.E.) and the *Āpastamba Sulba Sūtra*, composed by Āpastamba (600 B.C.E.), contained results similar to the *Boudhāyana Sulba Sūtra*.

1.3. Mathematical Activity in the Vedic Period

Records of Mathematical activity are mostly to be found in *Vedic* texts associated with ritual activities. However, as in many other early civilisations, the study of Arithmetic and Geometry was also impelled by secular considerations.

भागं हरेदवर्गान्नित्यं द्विगुणेन वर्गमूलेन ।
वर्गाद्वर्गे शुद्धे लब्धं स्थानान्तरे मूलम् ॥

Bhāgam haredavargānnityam dviguṇena vargamūlena |
Vargādvarge śuddhe labdham sthānāntare mūlam ||
Āryabhaṭīyam – 476 C.E.

One should divide the non–square place by twice the square root of the square place and then subtract the square from the next square place.

Arithmetic operations such as addition, subtraction, multiplication, fractions, squares, cubes and roots are enumerated in *Viṣṇu Purāṇa* attributed to *Vedavyās* (pre–1000 B.C.E.). Examples of Geometric knowledge (*rekhā–gaṇita*) are to be found in the *Śulva–Sūtras* of Boudhāyana (800 B.C.E.) and Āpastamba (600 B.C.E.), which describe techniques for construction of ritual altars in use during the *Vedic* era. It is likely that these texts tapped Geometric knowledge that might have been acquired much earlier, possibly in the Harappa period. Boudhāyana's *Sūtra* displays an understanding of basic Geometric shapes and techniques of converting one

Geometric shape (such as a rectangle) to another of equivalent (or multiple, or fractional) area (such as a square). While some of the formulations are approximations, others are accurate and reveal a certain degree of practical ingenuity as well as some theoretical understanding of basic Geometric principles. Modern methods of multiplication and addition probably emerged from the techniques described in *Śulva–Sūtras*.

Modern day commentators are divided on how some of the results were generated. Some believe that these results came about through trial and error – as rules of thumb, or as generalisations of observed examples. Others believe that once the scientific method came to be formalised in the *Nyāya Sūtras* – proofs for such results must have been provided, but these have either been lost or destroyed, or else were transmitted orally through the *Gurukula* system and only the final results were tabulated in the texts. In any case, the study of *Gaṇita* (Mathematics) was given considerable importance in the *Vedic* period. The *Vedānga Jyotiṣa* (1000 B.C.E.) observes: "Just as the feathers of a peacock and the jewel–stone of a snake are placed at the highest point of the body (at the crown/forehead), similarly, the position of *Gaṇita* is the highest amongst all branches of the *Vedas* and the *Śāstra*".

1.4. Pāṇini and Formal Scientific Notation

A particularly important development in the history of Indian Science that was to have a profound impact on all Mathematical treatises that followed was the pioneering work by Pāṇini (6th Century B.C.E.) in the field of Saṁskṛit grammar and linguistics. Besides expounding a comprehensive and scientific theory of phonetics, phonology and morphology, Pāṇiní provided formal production rules and definitions describing Saṁskṛit grammar in his treatise called *Aṣṭādhyāyi*. Basic elements such as vowels and consonants, parts of speech such as nouns and verbs were placed in classes. The construction of compound words and sentences was elaborated through ordered rules operating on underlying structures in a manner similar to formal language theory.

Today, Pāṇiní's constructions can also be seen as comparable to modern definitions of a Mathematical function. G G Joseph, in his book called "The Crest of the Peacock" argues that the Algebraic nature of Indian Mathematics arises as a consequence of the structure of the Saṁskṛit language.

Ingerman in his paper titled "Pāṇiní–Backus Form" finds Pāṇiní's notation to be equivalent in its power to that of Backus – Inventor of the Backus Normal Form used to describe the syntax of modern computer languages.

Thus Pāṇiní's work provided an example of a scientific notational model that could have propelled later Mathematicians to use abstract notations in characterising Algebraic equations and presenting Algebraic theorems and results in a scientific format.

1.5. Philosophy and Mathematics

Philosophical doctrines also had a profound influence on the developmer t of Mathematical concepts and formulations. Like the *Upaniṣads* (a research on ultimate truth – Supreme bliss) world view, space and time were considered limitless in Jain cosmology. This led to a deep interest in very large numbers and definitions of infinite numbers. Infinite numbers were created through recursive formulae, as in the *Anuyoga–dwāra–sūtra*. Jain Mathematicians recognised five different types of infinities: infinite –

- In one direction,
- In two directions,

- In area,
- Everywhere and
- Perpetually infinite.

Permutations and combinations are listed in the *Bhagvati–sūtras* (3rd Century B.C.E.) and *Sathananga–sūtra* (2nd Century B.C.E.).

Jain set theory probably arose in parallel with the *Syādvāda* system of Jain epistemology in which reality was described in terms of pairs of truth conditions and state changes. The *Anuyoga–dwāra–sūtra* demonstrates an understanding of the law of indices and uses it to develop the notion of logarithms. Terms like *Ardh Āched, Trik Āched* and *Chatur Āched* are used to denote log of base 2, log of base 3 and log of base 4 respectively. In *Satkhanḍagama* (*Shatkanḍakam*) various sets are operated upon by logarithmic functions of base two, by squaring and extracting square roots and by rising to finite or infinite powers. The operations are repeated to produce new sets. In other works the relation of the number of combinations to the co–efficient occurring in the binomial expansion is noted.

Since Jain epistemology allowed for a degree of indeterminacy in describing reality, it probably helped in grappling with indeterminate equations and finding numerical approximations to irrational numbers.

Buddhist literature also demonstrates an awareness of indeterminate and infinite numbers. Buddhist Mathematics was classified either as *Garna* (simple Mathematics) or *Saṅkhyān* (higher Mathematics). Numbers were deemed to be of three types: *Sankheya* (countable), *Asankheya* (uncountable) and *Anant* (infinite).

Philosophical formulations concerning *śūnya* – the emptiness or the void may have facilitated in the introduction of the concept of zero. While the zero (*bindu*) as an empty place holder in the place–value numeral system appears much earlier, Algebraic definitions of the zero and its relationship to Mathematical functions appear in the Mathematical treatises of Brahmagupta in the 7th Century C.E. Although there are different schools of thought about how early the symbol for zero came to be used in numeric notation in India, (Ifrah[2] arguing that the use of zero is already implied in Āryabhaṭa) tangible evidence for the use of the zero begins to proliferate towards the end of the Gupta period. Between the 7th Century C.E. and the 11th Century C.E., Indian numerals developed into their modern form alongwith the symbols denoting various Mathematical functions (such as plus, minus, square root, etc.), which eventually became the foundation stones of modern Mathematical notation.

1.6. The Indian Numeral System

> "We owe a lot to Indians, who taught us how to count, without which no worth–while scientific discovery could have been made".
>
> – Albert Einstein

To start with the representation of numbers were limited to three – all threesomes – three worlds, *trimoorthies, trikāl, triguṇas, threerāmas, three agnis, three Vyāhritis,* etc. Even *thulasi* and *vilva* leaves used during worship should be as threesome-s (*dalas*).

Later the representation of numbers was increased to Eight. That is the reason in Tamil this number was called as *ettu*. In Tamil *Ettā* means unable to reach. At this time everything was denoted in

[2] **Georges Ifrah** (called as Ifrah) is a French author and historian of Mathematics, especially numerals.

eight-some – *Ashta Lakshmee, Ashta Dik, Ashta Vasu*-s, *Ashta Gaja*-s, *Ashta Aśwa*-s, *Ashtamāsiddhies*, etc.

Also in Saṁskṛit the number nine was called as *nava*, which means new. The same is the case in Roman numbers also – number nine was not denoted as VIIII instead it was mentioned as IX – one less than ten. In Hindi language number nineteen is *unnis* (one less than twenty), *unthees* (one less than thirty), etc.

Similar is the case of Saṁskṛit language also - number nineteen is *ekonavimshat* (one less than twenty), *ekonatrimshat* (one less than thirty), etc.

Again in Tamil language number –

nine is *onpathu* – one short of ten,
number ninety is – not *thonpathu*– it is
thonooru – ten short of hundred
number nine hundred is – not *onpathu nooru* – it is
thollāyiram – hundred short of thousand

Although the Chinese were also using a decimal based counting system, they lacked a formal notational system whose abstraction and elegance was not comparable to that of their Indian counterpart and it was the Indian notational system that reached the Western world through the Arabs and has now been accepted universally. Hence it came to be called as Arabic numerals by the Westerners, which has come to stay, though it is Indian Numeral indeed.

Several factors contributed to this development whose significance is perhaps best stated by French Mathematician, Laplace as:

"The ingenious method of expressing every possible number using a set of ten symbols (each symbol having a place value and an absolute value) emerged in India. The idea seems so simple nowadays that its significance and profound importance is no longer appreciated. Its simplicity lies in the way it facilitated calculation and placed arithmetic foremost amongst useful inventions".

Brilliant as it was, this invention was no accident. In the Western world, the cumbersome Roman numeral system posed as a major obstacle and in China the pictorial script posed as a hindrance. But in India, almost everything was in place to favour such a development. There was already a long and established history in the use of decimal numbers and philosophical and cosmological constructs encouraged a creative and expansive approach to number theory.

Evolution of Numerals

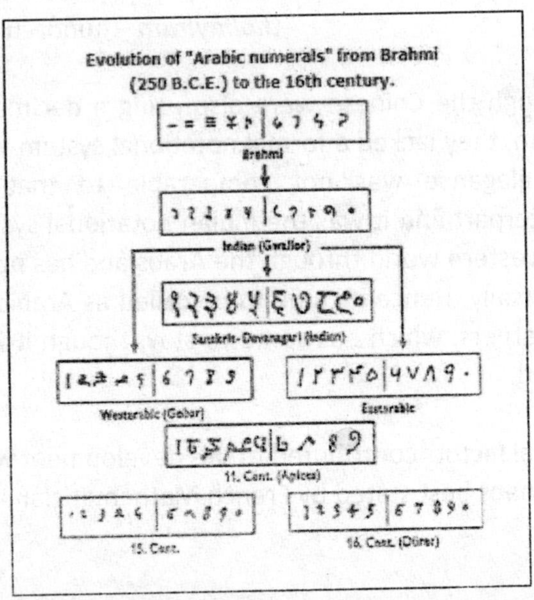

Modern (current) Numerals	1	2	3	4	5	6	7	8	9	0
Eastern Arabic–Indic (Persian and Urdu)	۱	۲	۳	۴	۵	۶	۷	۸	۹	۰
Devanāgari (1100 C.E.)	१	२	३	४	५	६	७	८	९	०
Brāhmi (300 B.C.E.)	𑁒	𑁓	𑁔	𑁕	𑁖	𑁗	𑁘	𑁙	𑁚	

1.7. *Siddhāntic* and Classical Age

In the period 5th Century B.C.E., Indian *Siddhānta* works based on theoretical Astronomical principles were composed. Traditional number of such *Siddhāntas* is 18 of which the *Pitāmaha Siddhānta* is the oldest. According to *Nārada Purāṇa* (chapter 54, verses 1–2), Indian *jyotiṣa śāstra* had 4 lakhs verses. One of the significant achievements of this period was the Indian invention of sine function, the development of Trigonometry and its use in Astronomy. In fact the Trigonometry based on Sine (instead of chord) and related functions is another important gift of India to the world of Mathematics (the foremost gift being decimal place – value system of numerals).

The Indian sine of an arc in circle of radius 'R' is equal to the chord of twice the arc and in value is equal to Rsinθ, where sinθ is the modern sine of the angle subtended by the given arc at the centre of the circle. Like the Indian numerals the use of Indian Trigonometry based on the Sines also had spread to the whole world, especially to the European countries through the Arabs. The Arabic translation of the Saṁskṛit Astronomical work was called *Sindhind* (Indian Planetary System) from which descended a long tradition within Islamic Astronomy extending upto Spain, for several centuries. In fact the English word 'Sine' itself is derived from the original Saṁskṛit word *jyā* through the corrupted Arabic transliterations 'Jeeb' and 'Jayab', which were translated into Latin as 'sinus'.

1.8. The Decimal System in Harappa

In India a decimal system was already in place during the Harappan period, as indicated by an analysis of Harappa weights and measures. Weights corresponding to ratios of 0.05, 0.1, 0.2, 0.5, 1, 2, 5, 10, 20, 50, 100, 200 and 500 have been identified, as have scales with decimal divisions. A particularly notable characteristic of Harappa weights and measures is their remarkable accuracy. A bronze rod marked in units of 0.367 inches points to the degree of precision demanded in those times. Such scales were particularly important in ensuring proper implementation of town planning rules that required roads of fixed widths to run at right angles to each other, for drains to be constructed of precise measurements and for homes to be constructed according to specified guidelines. The existence of a graded system of accurately marked weights points to the high–level of development of trade and commerce in Harappa society.

1.9. The Spread of Indian Mathematics

The study of Mathematics appeared to slow down after the onslaught of the Islamic invasions and the conversion of colleges and universities to 'Madrasahs'. But this was also the time when Indian Mathematical texts were increasingly being translated into Arabic and Persian. Although Arab scholars relied on a variety of sources including Babylonian, Syrian, Greek and some Chinese texts, Indian Mathematical texts played a particularly important role. Scholars such as Ibn Tariq and Al–Fazari (8th Century C.E., Baghdad), Al–Kindi (9th Century C.E., Basra), Al–Khwarizmi (9th

Century C.E., Khiva), Al–Qayarawani (9th Century C.E., Maghreb, author of "Kitab fi al–hisab al–hindi"), Al–Uqlidisi (10th Century C.E., Damascus, author of "The book of Chapters in Indian Arithmetic"), Ibn–Sina (Avicenna), Ibn al–Samh (Granada, 11th Century C.E., Spain), Al–Nasawi (Khurasan, 11th Century C.E., Persia), Al–Beruni (11th Century C.E., born Khiva, died Afghanistan), Al–Razi (Teheran) and Ibn–Al–Saffar (11th Century C.E., Cordoba) were amongst the many who based their own scientific texts on translations of Indian treatises. Records of the Indian origin of many proofs, concepts and formulations were obscured in the later centuries. However, the enormous contributions of Indian Mathematics were generously acknowledged by several important Arabic and Persian scholars, especially in Spain. Abbasid scholar Al–Gaheth wrote: "India is the source of knowledge, thought and insight". Al–Maoudi (956 C.E.), who travelled Western India, wrote about the greatness of Indian Science. Said–Al–Andalusi (11th Century C.E.), a Spanish scholar was amongst the most enthusiastic in his praise of Indian civilisation and specially remarked on Indian achievements in Science and Mathematics. Of course, eventually, Indian Algebra and Trigonometry reached Europe through a cycle of translations, travelling from the Arab world to Spain and Sicily, finally penetrating all of Europe. At the same time, Arabic and Persian translations of Greek and Egyptian scientific texts became more readily available in India.

1.10. The Kerala School

Although, it appears, that original works in Mathematics ceased in much of Northern India after the Islamic conquests, Benares survived as a centre for Mathematical study and an important school of Mathematics blossomed in Kerala. Mādhava (1380–90 – 14th Century C.E., Kochi) made important Mathematical discoveries that were found out by European Mathematicians two centuries later. His period can very well be called as Golden Years of Indian Mathematics. His series expansion of the Cosine and Sine functions were earlier than that of Newton's by almost three centuries. Historians of Mathematics, Rajagopal, Rangachari and Joseph considered his contributions instrumental in taking Mathematics to the next stage of modern classical analysis. Nīlakanṭha (15th Century C.E., Tiruchur, Kerala) extended and elaborated upon the results of Mādhava while Jyeṣṭadeva (16th Century C.E., Kerala) provided detailed proofs of the theorems and derivations of the rules contained in the works of Mādhava and Nīlakanṭha. It is also notable that Jyeṣṭadeva's Yuktibhāṣā, which contained commentaries on Nīlakanṭha's *'Tantrasamgraha'* included elaborations on planetary theory later adopted by Tycho Brahe and Mathematics that anticipated work by later Europeans. It has been discussed; debated and proved that *Yuktibhāṣā* is actually the basis for the present day Calculus based Mathematics. Citrabhānu (16th Century C.E., Kerala) gave integer solutions to twenty–one types of systems of two Algebraic equations, using both Algebraic and Geometric methods in developing his results. Important discoveries by the Kerala Mathematicians included the Newton–Gauss interpolation formula, the formula for the sum of an infinite series and a series notation for π (*pi*). Charles Whish (1835, published in the "Transactions of the Royal Asiatic Society of Great Britain and Ireland") was one of the first Westerners to recognise that the Kerala school had anticipated by almost 300 years before itself many European developments in that field.

Yet, very few modern compendiums on the history of Mathematics have paid adequate attention to the often pioneering and revolutionary contributions of early Indian Mathematicians. But, a significant body of Mathematical works were produced in the Indian subcontinent. The science of Mathematics played a pivotal role not only in the industrial revolution but in the scientific developments that have occurred since. No other branch of science is complete without Mathematics. Not only did India provide the financial capital for the industrial revolution, but also, it provided vital elements of the scientific foundation without which humanity could not have entered this modern age of advanced science and high technology.

Transmission of the Indian Numeral System: Evidence for the transmission of the Indian Numeral System to the West is provided by Joseph ("Crest of the Peacock"):

- Quotes Severus Sebokht (662) in a Syriac text describing the "subtle discoveries" of Indian Astronomers as being "more ingenious than those of the Greeks and the Babylonians" and "their valuable methods of computation which surpass description" and then goes on to mention the use of nine numerals.
- Quotes from "Liber abaci" (Book of the Abacus) by Fibonacci (1170–1250): the nine Indian numerals are ... with these nine and with the sign '0' which in Arabic is *sifr*, any desired number can be written. (Fibonacci learnt about Indian numerals from his Arab teachers in North Africa).

Influence of the Kerala School: Joseph ("Crest of the Peacock") suggests that Indian Mathematical manuscripts may have been brought to Europe by Jesuit priests such as Mateo Ricci who spent two years in Kochi (Cochin) after being in Goa in 1580 C.E. Kochi is only 70 km from Thrissur (Trichur), which was then the largest repository of Astronomical documents. Whish and Hyne – two European Mathematicians obtained their copies of works by the Kerala Mathematicians from Thrissur and it is not inconceivable that Jesuit monks may have also taken copies to Pisa (where Galileo, Cavalieri and Wallis spent time), or Padau (where James

Gregory studied) or Paris (where Mersenne who was in touch with Fermat and Pascal, acted as an agent for the transmission of Mathematical ideas).

1.11. The sign for Zero

The history of numerals holds a dual fascination since development of science presupposes numbers and the usage of numbers by the common folk can reveal surprising aspects of cultural life. Many of these dual issues for different civilisations are explored in Karl Menninger's "Number words and Number Systems" (1958). But the story of how number systems evolved in early times is not very clear. Perhaps this evolution can be best seen in Indian civilisation, which since the *Ṛg Vedic* age seems to have been interested in numeric questions. Georges Ifrah in his "From One to Zero: A Universal History of Numbers" has given an interesting account of the history of the Indian numerals. These are the common numerals and they are also called the Hindu–Arabic numerals. Ifrah has sketched a plausible explanation for how the place value system of the Indians may have arisen upon the use of the counting boards. The place value system with a clear use of zero goes back at least to 458 C.E., where it is used in *Lokavibhāga*, a Jain work on cosmology. The earliest epigraphical evidence relating to the use of the nine numerals in a place value manner goes back to 595 C.E., on a copperplate deed from *Sāṅkheda*. But some symbol for zero must have been in use in India as early as 200 B.C.E., since Piṅgala in his *Chandas–śāstra* describes a representation system that is binary, as in modern computers that would clearly need a symbol for zero. And much before Piṅgala the grammarian Pāṇini used the notion of the null (zero) operation in his *Aṣṭādhyāyī*.

One question that has not been addressed by Ifrah or earlier historians is the shape of the zero sign. Why is it in the form of a circle or an oval rather than some other shape? The evolution of the shapes for the other signs is well understood, but not for zero. However, it has generally been assumed that the sign for zero appeared suddenly from nowhere. It is attempted to sketch a developmental process that is likely to have led to the round form of the zero sign. This allows us to be more definite about the epoch when the sign was developed. In this process we can also consider why the zero sign in India came to be fully developed with both the place value notion as well as the notion of null value associated with it.

Numbers in Saṁskṛit have been expressed in different orders. Numerals were expressed in symbolic form in the place notation in the order starting from the units. This has been mentioned both in *Lokavibhāga* and *Āryabhaṭīya* as well. This representation is the order in which numbers are expressed in classical Saṁskṛit. Thus 14 (*caturdaśa* 4 + 10) and 517 (*saptadaśādhikapańchaśatam* – 7 + 10 + 500) have been represented. While this is how upto numbers 99 are built up in Indo–Aryan languages, the numbers above 100 were actually represented in the reverse order excepting for the last two places. This is also true for Saṁskṛit as confirmed by checking numerical references in *Śulva sūtra–s*. Thus in *Boudhāyana*, 225 is expressed as *dve śate pańchavimśatiśca* i.e. 200 + 5 + 20 or 187 is written as 100 + 7 + 80. Likewise *Vedānga Jyotiṣa* expresses the numbers, sometimes, in the same style.

In other words there existed two styles for expressing numbers that were well established:

1. In formal texts (to be consistent with the structure of numerals) until 99 all numbers were expressed starting with units;
2. Actual writing in terms of numeral signs was in the usual place value form starting from the highest order.

This is confirmed by epigraphical evidence (Datta and Singh 1935) as well, implied by rules for various Mathematical operations in *Āryabhaṭīya*, *Brāma–Sphuṭa–Siddhānta* and other Mathematical works.

Brāhmi script was in use, in general, in India during which period the sign for zero is likely to have been developed. The number 10 in Brāhmi, before the advent of zero was written as a fish sign  . In later forms it was also written as a single curving stroke , or with the Asokan 1 attached to a circle  .

Using these forms the numbers such as 15 were written as 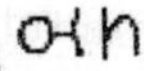 or 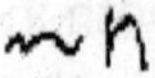 where  is 5 in Brāhmi. Therefore it is not unlikely that the shape of zero was determined by the oval related to the fish sign of number 10 in Brāhmi. In such numbers the zero sign clearly had the null (*śūnya*) value, which explains its name. It can also be seen how the two concepts expressed by the Indian zero, namely those of the place value and that of nothing, are likely to have become evident. Emptiness was never visualised in Māyan or Greek or any other numbering system. Though they all felt the need for zero they all were struggling a lot in this regard. It is the Indian Philosophy and verses, which could perceive emptiness and infinity and hence could bring a notation for zero.

The significance of the reversed order usage may have played a role in the clarification of the concept of zero as 'null'. An addition such as 10 + 3 would now be performed as:

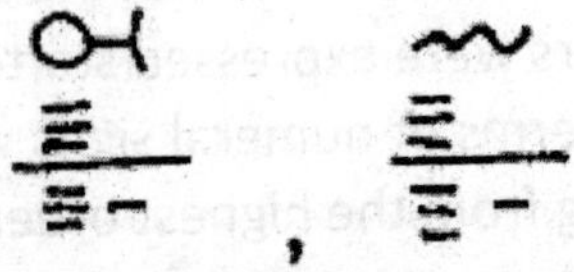

and here also zero would correspond to the value 'nothing'.

Perhaps the simultaneous existence of the two forms of expressing numbers helped in the development of the dual concepts associated with the zero sign.

We also encounter the  form for 10 in the first and second Century C.E. in Nasik inscriptions and in the first and third Century C.E., in Andhra and Kṣatrapa inscriptions respectively.

The curved form  is seen in the 4th Century C.E. in Jaggayapeta and Pallava grants (Datta and Singh, 1938). Therefore it can be construed that the development of the zero sign occurred in these epochs.

But these epochs do not provide a definite era for the discovery of the zero sign, since it is likely that the new usage competed with the traditional number system for centuries. In fact one would expect that inscriptions and deed–plates would tend to follow the older and more commonly known style for a longer time. For a parallel, it can be considered that in Europe it took the Indian numerals more than 600 years to oust the less convenient Roman numerals. Even in India the older additive system with special signs for 10, 20, 30, etc., continued to be used, alongside the place value system, for centuries. (In Information Technology, such a usage is called parallel testing).

The development of the zero sign in India was motivated by numerical calculations. This is to be contrasted from the zero signs across Babylon and Mexico, where it was necessitated by Astronomy and calendric calculations. The Babylonian Astronomical tablets use a sexagesimal numeration system. But it is imperfectly developed being partly additive and partly place valued as within the base of 60, a decimal system is used. The Babylonian system has only three specific symbols, namely those for 1, 10 and the later symbol for zero. Unless the groups of wedge marks are separated it is always possible to miscalculate the indicated number. For example:

can be interpreted either as 2; 0; 25 = 7225 or as 2; 0; 20; 5 = 217205.

The Mayans, on the other hand, used a vi–gesimal (base 20) system but with a serious irregularity since its units were 1, 20, 18 x 20, 28 x 20^2, 18 x 20^3 and so on. Thus in this system the glyph representing a seashell (which is 0) does not work as an operator, as it should in a true place value system. Furthermore the numbers upto 20 are additive as in the case of the Babylonian system and therefore there exists the same possibility of ambiguity.

The surviving inscriptions and codices do not write the numbers without specifying the units, which eliminates ambiguity but shows that the abstract nature of the place value number system was not fully understood. Clearly this system was also not designed for the needs of ordinary calculations. The rationale behind the Mayan system was the counting of the days of 18 months, each of 20 days.

In view of the demonstration of the derivation of the Brāhmi numerals from the earlier Indus numerals, especially how the Indus 5 and 10 became the corresponding Brāhmi signs, the above derivation pushes the ancestry of the zero sign to the third millennium B.C.E.

The choice of a circle, or sky, or space *kha* for zero had a happy reinforcement from the Brāhmi sign for *kha*, which is a circle with a hook on top of it.

That means the development of the concept of zero and its form in India were related to the form of the Brāhmi sign for 10. Furthermore we see that the explication of the full power of the symbol may have been aided by the two different ways of expressing numbers in Saṁskṛit. We have identified a late form of Brāhmi 10 as the immediate parent of the zero sign.

1.12. Zero and the Place Value System

Let us read through a story. Let us imagine that we are living in the age of the early man. There are no numbers. Suppose all that we know of counting are the following symbols:

@ – represents 'zero'
# – represents 'one'
$ – represents 'two'

One day, some apples are collected and scattered down under an apple tree. If we are required to express the quantity of apples using the notation available (listed above), how would it be done? If there were 8 apples, how the notation be used? The apples would have been placed into groups of three, say like this:

Here, two groups are complete (3 in a group) and one is partial with 2 apples. Now, to represent the number of apples, let us first write the number of total groups, followed by the number of apples in the partial group, like this – $$. Therefore, '$$' successfully tells that there are 8 apples.

If there were 14 apples, how would it be expressed?

Here, the number of total groups = 5 and the number of apples in the partial group = 2. The number '4' can be represented as '##' (this is nothing but grouping the groups again into one total and a partial group with one apple).

Hence, 14 can be represented as '##$' (4 groups + 2 in the partial group).

If there were 27 apples, how would the representation look like? To what level of grouping of the groups does it lead to? The grouping is very difficult. Here comes the role of zero in such a representation.

Actually, the answer is ' #@@@'. It can be observed that in the process of grouping, not once did we get a partial group? How should such absence of partial groups be represented? Obviously, we need some notation to denote absence. That is zero. Hence the discovery of zero by the Indian Mathematicians can be appreciated.

Far more important to the development of modern Mathematics than either Greek or Indian Geometry was the development of the place value system of enumeration, the base ten system of calculation, which uses nine numerals and zero to represent numbers ranging from the most minuscule decimal to the most inconceivably large power of ten. This system of enumeration was not developed by the Greeks, whose largest unit of enumeration was the myriad (10,000) or in China, where also 10,000 was the largest unit of enumeration until recent times. Nor was it developed by the Arabs, despite the fact that this numeral system is commonly called as Arabic numerals in Europe, where this system was first introduced by the Arabs in the 13th Century C.E.

Rather, this system was invented in India, where it is evident of its ancient origin. The *Yajur Veda Samhitā*, one of the *Vedic* texts predating Euclid and the Greek Mathematicians by at least a millennium, lists names for each of the units of ten up to 10^{12} (says Subbarayappa B V in his book titled “India’s Contributions to the History of Science”). Later Buddhist and Jain authors extended this list as high as the 10^{53}, far exceeding their Greek contemporaries, who, lacking a system of enumeration, were unable to develop abstract Mathematical concepts. The largest number known to ancient Greeks was 10^3 (Myriad) and to the Romans 10^4 (Mille). In India, from the time of *Vedas*, both very large as well as very small numbers were computed.

The place value system of enumeration is in fact built into the Saṁskṛit language, where each power of ten is given a distinct name. Only the unit, ten, hundred and thousand are named differently in English and ten thousand, hundred thousand, ten million, hundred million and so forth makes use of auxiliary bases such as thousand, million, etc.

Ifrah has commented that by giving an individual name to each power of ten, the Saṁskṛit system gave no special importance to any number. Thus the Saṁskṛit system is obviously superior to that of the Arabs (for whom the thousand was the limit), or the Greeks and Chinese (whose limit was ten thousand) and even to English system (where the names thousand, million etc., continue to act as auxiliary bases). Instead of naming the numbers in groups of three, four or eight orders of units, the Indians, from a very early date, expressed them taking the powers of ten and the names of the first nine units individually. In other words, to express a given number, one had to only place the name indicating the order of units immediately below it and the one immediately above it. That is exactly what is required in order to gain a precise idea of the place–value system, the rule being presented in a natural way and thus appearing self–explanatory. To put it in a nutshell, the Saṁskṛit numeral system contained the very key/ formed the crux for the discovery of the place–value system.

As Ifrah has shown at length, there is little doubt that the place–value numeral system was developed in India and this system is embedded in the Saṁskṛit language, several aspects of which make it a very logical language, well suited for Scientific and Mathematical reasoning. Nor did this system exhaust Indian ingenuity; as Van Nooten has shown, Piṇgalā, who lived in the first Century B.C.E., developed a system of binary enumeration convertible to decimal numerals, described in his *Chandas–śāstra*. His system is quite similar to that of Leibniz, who lived roughly fourteen hundred years later.

India is also the focus of another closely related and equally important Mathematical discovery, the numeral zero. The oldest known text to use zero is a Jain text entitled the *Lokavibhāga*, which has been definitely dated to Monday, 25th August 458 C.E., (Ifrah 2000:417–1 9). This concept, combined by the place–value system of enumeration, became the basis for a classical era of renaissance in Indian Mathematics.

The Indian numeral system along with its place value and decimal system of enumeration came to the attention of the Arabs in the seventh or eighth Century C.E. and served as the basis for the well–known advancement in Arab Mathematics, represented by figures such as al–Khwarizmi. It reached Europe in the 12th Century C.E., when Adelard of Bath translated al–Khwarizmi's works into Latin. (Subbarayappa 1970:49). But the Europeans were at first resistant to this system, being attached to the far less logical roman numerals. But their eventual adoption of this system led to the scientific revolution that began to sweep Europe beginning in the 13th Century C.E.

1.13. India and the Scientific Revolution

Dr. David Gray in his book Indic Mathematics writes: the study of Mathematics in the West has long been characterised by a certain ethnocentric bias, a bias which most often manifests not in explicit racism, but in a tendency towards undermining or eliding the real contributions made by non–Western civilisations. The debt owed by the West to other civilisations and to India in particular, goes back to the earliest epoch of the 'Western' scientific tradition, the age of the classical Greeks and continued up until the dawn of the modern era, the renaissance, when Europe was awakening from its dark ages.

Dr. Gray goes on to list some of the most important developments in the history of Mathematics that took place in India, summarising the contributions of luminaries such as Āryabhaṭa, Brahmagupta, Mahāvīra, Bhāskara and Mādhava. He avers that "the role played by India in the development of the scientific revolution in Europe is no mere footnote – easily and inconsequentially swept under the rug of Eurocentric bias. To do so is to distort history and to deny India one of its greatest contributions to world civilisation".

George Ghevarughese Joseph, in an important article, titled "Foundations of Euro–centrism in Mathematics", argued that "the standard treatment of the history of non–European Mathematics is a product of historiographical bias (conscious or otherwise) in the selection and interpretation of facts, which, as a consequence, results in ignoring, devaluing or distorting contributions arising outside European Mathematical traditions". (1987:14)

Due to the legacy of colonialism, the exploitation of which was ideologically justified through a doctrine of racial superiority, the contributions of non–European civilisations were often ignored, or, as Joseph argued, even distorted, in that they were often misattributed as European, i.e., when non–European contributions were so great as to resist such treatment, they were typically devalued, considered inferior or irrelevant to Western Mathematical traditions.

This tendency has not only led to the devaluation of non–Western Mathematical traditions, but has distorted the history of Western Mathematics as well. This has led, as Sabetai Unguru has argued, towards a tendency to read more advanced Mathematical concepts into the relatively simplistic Geometrical formulations of Greek Mathematicians such as Euclid, despite the fact that the Greeks lacked not only Mathematical notation, but even the place–value system of enumeration, without which advanced Mathematical calculation is impossible. Such ethnocentric revisionist history resulted in the attribution of more advanced Algebraic concepts, which were actually introduced to Europe over a millennium later by the Arabs, to the Greeks. And while the contributions of the Greeks to Mathematics was quite significant, the tendency of some Math historians to jump from the Greeks to renaissance Europe results not only in an ethnocentric history, but an inadequate history as well, one which fails to take into account the full history of the development of modern Mathematics, which is by no means a purely European development.

Mathematics and Music: Piṅgalā (3rd Century C.E.), author of *Chandas–śāstra* explored the relationship between combinatory and Musical theory anticipating Mersenne (1588–1648) author of a classic on musical theory.

Mathematics and Architecture: Interest in Arithmetic and Geometric series may have also been stimulated by (and influenced) Indian Architectural designs – (as in temple *Shikaras, Gopurams* and corbelled temple ceilings). Of course, the relationship between Geometry and architectural decoration was developed to its greatest heights by Central Asian, Persian, Turkish, Arab and Indian architects in a variety of monuments commissioned by the Islamic rulers.

1.14. Emergence of Calculus

In the course of developing a precise mapping of the lunar eclipse, Āryabhaṭa was obliged to introduce the concept of infinitesimals – i.e. *tātkālika gati* to designate the infinitesimal, or near instantaneous motion of the moon and express it in the form of a basic differential equation. Āryabhaṭa's equations were elaborated on by Manjula (10th Century C.E.) and Bhāskarācārya (12th Century C.E.) who derived the differential of the sine function. Later Mathematicians used their intuitive understanding of integration in deriving the areas of curved surfaces and the volumes enclosed by them.

1.15. Importance of Astronomy

The growth of trade and commerce, particularly lending and borrowing demanded an understanding of both simple and compound interest which probably stimulated the interest in arithmetic and Geometric series. Brahmagupta's description of negative numbers as debts and positive numbers as fortunes points to a link between trade and Mathematical study. Knowledge of Astronomy – particularly knowledge of the tides and the stars was of great importance to trading communities who crossed oceans or deserts at night. This is borne out by numerous references in the *Jātaka* tales and several other folk–tales. Any young person who wished to embark on a commercial venture was inevitably required to first gain some base in Astronomy. This led to a proliferation of teachers of Astronomy, who in turn received training at Universities such as at Kusumpura (Bihar) or Ujjain (Central India) or at smaller local colleges or *Gurukuls.* This also led to the exchange of texts on Astronomy and Mathematics amongst scholars and the transmission of knowledge from one part of India to another. Virtually every Indian state produced great Mathematicians who wrote commentaries on the works of other Mathematicians (who may have lived and worked in a different part of India many centuries earlier). Saṁskṛit served as the common medium of scientific communication.

The science of Astronomy was also spurred by the need to have accurate calendars and a better understanding of climate and rainfall patterns for timely sowing and for choosing the right crops. At the same time, religion and astrology also played a role in creating an interest in Astronomy and negative fallout of this irrational influence was the rejection of scientific theories that were far ahead of their time. One of the greatest scientists of the Gupta period – Āryabhaṭa (born in 476 C.E., Kusumpura, Bihar) provided a systematic treatment of the position of the planets in space. He correctly calculated the axial rotation of the earth and inferred correctly that the orbits of the planets were **ellipses**. He also correctly deduced that the moon and the planets shined by **reflected sunlight** and provided a valid explanation for the **solar and lunar eclipses** rejecting the superstitions and mythical belief systems surrounding the phenomenon. Although Bhāskara I (Saurashtra, 6th Century C.E. and follower of the Asmaka School of Science, Nizamabad, Andhra Pradesh) recognised his genius and the tremendous value of his scientific contributions, some later Astronomers continued to believe in a static earth and rejected his rational explanations of the eclipses. But in spite of such setbacks, Āryabhaṭa had a profound influence on the Astronomers and Mathematicians who followed him, particularly those from the *Asmaka* School.

Mathematics played a vital role in Āryabhaṭā's revolutionary understanding of the solar system. His value of π (pi), in calculating the circumference of the earth (62832 miles) and the length of the solar year were remarkably close approximations (within about 13 minutes) to current calculations. In making such calculations, Āryabhaṭa had to solve several Mathematical problems that had not been addressed before including problems in Algebra (*bīja gaṇit*) and Trigonometry (*trikoṇamiti*).

Bhāskara I continued where Āryabhaṭa left off and discussed in further detail topics such as the longitudes of the planets; conjunctions of the planets with each other and with bright stars; risings and settings of the planets; and the lunar crescent. Again, these studies required an even more advanced Mathematics and in the course of developing a precise mapping of the lunar eclipse, Āryabhaṭa was obliged to introduce the concept of infinitesimals – i.e. *tātkālika gati* to designate the infinitesimal, or near instantaneous motion of the moon and express it in the form of a basic differential equation. Amongst his most important contributions, the formula for calculating the Sine function, which had an accuracy of 99%, was the best. He also did pioneering work on indeterminate equations and considered for the first time quadrilaterals with all the four sides unequal and none of the opposite sides parallel.

Another important Astronomer–Mathematician was Varāhamihira (6th Century C.E., Ujjain) who compiled previously written texts on Astronomy and made important additions to Āryabhaṭa's Trigonometric formulae. His works on permutations and combinations complemented what had been previously achieved by Jain Mathematicians and provided a method of calculation of *nCr* that closely resembles the much more recent Pascal's Triangle. In the 7th Century C.E., Brahmagupta did important work in enumerating the basic principles of Algebra. In addition to listing the Algebraic properties of zero, he also listed the Algebraic properties of negative numbers. His work on solutions to quadratic indeterminate equations anticipated the work of Euler and Lagrange.

1.16. Applied Mathematics, Solutions to Practical Problems

Developments also took place in applied Mathematics such as in creation of Trigonometric tables and measurement units. Yativrsabha's work *'Tiloyapaṇatti'* (6th Century C.E.) gives various units for measuring distances and time and also describes the system of infinite time measures.

In the 9th Century C.E., Mahāvīrācārya (Mysore) wrote "Gaṇita–sārasangraha" where he described the currently used method of calculating the Least Common Multiple (LCM) of given numbers. He also derived formulae to calculate the area of an ellipse and a quadrilateral inscribed within a circle (something that had also been looked at by Brahmagupta). The solution of indeterminate equations also drew considerable interest in the 9th Century C.E. and several Mathematicians contributed approximations and solutions to different types of indeterminate equations.

Mathematical investigation continued into the 10th Century C.E. – Vijayanandi (of Benares, whose *'Karaṇatilaka'* was translated by Al–

Beruni into Arabic) and Śripati of Maharashtra are amongst the prominent Mathematicians of the 10th Century C.E.

The leading light of 12th Century C.E., in Indian Mathematics was Bhāskarācārya who came from a long–line of Mathematicians and was head of the Astronomical observatory at Ujjain. He left several important Mathematical texts including the *Leelāvati* and '*Bījagaṇita*' and the "*Siddhānta–śiromani*", an Astronomical text. He was the first to recognise that certain types of quadratic equations could have two solutions. His *Cakravāla* method of solving indeterminate equations preceded European solutions by several centuries and in his "*Siddhānta–śiromani*" he postulated that the earth had a gravitational force and broached the fields of infinitesimal calculation and integration. In the second part of his treatise, there are several chapters relating to the study of the sphere and its properties and applications to geography, planetary mean motion, eccentric epicyclical model of the planets, first visibilities of the planets, the seasons, the lunar crescent, etc. He also discussed Astronomical instruments and spherical Trigonometry. Of particular interest are his Trigonometric equations:

Sin(a + b) = Sina Cosb + Cosa Sinb; and
Sin(a – b) = Sina Cosb – Cosa Sinb;

आकृष्टशक्तिश्च मही तया खस्थं गुरु स्वाभिमुखं स्वशक्त्या ।
आकृष्यते तत्पततीव भाति समे समन्तात् क्व पतत्वियम् खे ॥

ākṛṣṭaśaktiśca mahī tayā khastham guru svābhimukham svaśaktyā |
ākṛṣyate tatpatatīva bhāti same samantāt kva patatviyam khe ||

Siddhānta–śiromani – Bhuvanakośa: – 6

Massive celestial bodies are attracted powerfully towards the Earth by her own (gravitational) force and they appear to fall as a result of such attraction, but when equal forces act on a body in space from all sides where will it fall?

The universal law of gravitation was propounded by Sir Isaac Newton in the 17th Century C.E., much later than the above.

*****

2. Classical Indian Mathematicians

There was many a Mathematician in our country, who has contributed a lot to the Mathematical world. It is unfortunate that all their works available are not available now. However, some of the major contributors and their contributions are worth noting.

The Vedic religion with its sacrificial rites began to wane and other religions began to replace it. One of these was Jainism, a religion and philosophy, which was founded in India around the 6th Century B.C.E. Although the period after the decline of the Vedic religion up to the time of Aryabhata I around 500 C.E. used to be considered as a dark period in Indian Mathematics, recently it has been recognised as a time when many Mathematical ideas were considered. In fact Aryabhata is now thought of as summarising the Mathematical developments of the Jaina as well as beginning of the next phase.

The main topics of Jaina Mathematics in around 150 B.C.E. were:

- The theory of numbers
- Arithmetical operations
- Geometry
- Operations with fractions
- Simple equations
- Cubic equations
- Quartic equations and
- Permutations & combinations.

More surprisingly the Jaina developed a theory of the infinite containing different levels of infinity, a primitive understanding of indices and some notion of logarithms to base 2. One of the difficult problems facing historians of Mathematics is deciding on the date of the Bakhshali manuscript. If this is a work which is indeed from 400 C.E., or at any rate a copy of a work which was originally written at this time, then the understanding of the achievements of Jaina Mathematics will be greatly enhanced.

If the Vedic religion gave rise to a study of Mathematics for constructing sacrificial altars, then it was Jaina cosmology which led to ideas of the infinite in Jaina Mathematics. Later Mathematical advances were often driven by the study of Astronomy. Well perhaps it would be more accurate to say that Astronomy formed the driving force since it was that 'Science', which required accurate information about the planets and other heavenly bodies and so, encouraged the development of Mathematics. Religion too played a major role in astronomical investigations in India for accurate calendars had to be prepared to allow religious observances to occur at the correct times. Mathematics then was still an applied science in India for many centuries with Mathematicians developing methods to solve practical problems.

Yavanesvara, in the 2nd Century C.E., played an important role in popularising Astrology when he translated a Greek astrology text dating from 120 B.C.E. If he had made a literal translation it is doubtful whether it would have been of interest to more than a few academically minded people. He popularised the text, however, by resetting the whole work into Indian culture using Hindu images with the Indian caste system integrated into the text.

By about 500 C.E. the classical era of Indian Mathematics began with the work of Aryabhata. His work was both a summary of Jaina Mathematics and the beginning of new era for Astronomy and Mathematics. His ideas of Astronomy were truly remarkable. He replaced the two demons Rāhu, the *Dhṛva* Rāhu which causes the phases of the Moon and the Parva Rāhu which causes an eclipse by covering the Moon or Sun or their light, with a modern theory of eclipses. He introduced Trigonometry in order to make his astronomical calculations, based on the Greek epicycle theory and he solved with integer solutions indeterminate equations which arose in Astronomical theories.

Āryabhata headed a research centre for Mathematics and Astronomy at Kusumapura in the Northeast of the Indian subcontinent. A school to study his ideas grew up there but more than that, Aryabhata set the agenda for Mathematical and Astronomical research in India for many centuries to come. Another Mathematical and Astronomical centre was at Ujjain, also in the North of the Indian subcontinent, which grew up around the same time as Kusumapura. The most important of the Mathematicians at this second centre was Varāhamihira who also made important contributions to Astronomy and Trigonometry.

The main ideas of Jaina Mathematics, particularly those relating to its cosmology with its passion for large finite numbers and infinite numbers, continued to flourish with scholars such as Yativrsabha. He was a contemporary of Varāhamihira and of the slightly older Aryabhata. It should also be noted that the two schools at Kusumapura and Ujjain were involved in the continuing developments of the numerals and of place-valued number systems. The next figure of major importance at the Ujjain school was Brahmagupta near the beginning of the seventh Century C.E. and he had made one of the most major contributions to the development of the numbers systems with his remarkable contributions on negative numbers and zero. It is a sobering thought that eight hundred years later European Mathematics would be struggling to cope without the use of negative numbers and of zero.

These were certainly not Brahmagupta's only contributions to Mathematics. Far from it, he made other major contributions in to the understanding of integer solutions to indeterminate equations and to interpolation formulae invented to aid the computation of Sine tables.

The Hindu astronomers did not possess a general method for solving problems in spherical Astronomy, unlike the Greeks who systematically followed the method of Ptolemy, based on the well-known theorem of Menelaus. But, by means of suitable constructions within the armillary sphere, they were able to reduce many of their problems to comparison of similar right-angled plane triangles. In addition to this device, they sometimes also used the theory of Quadratic Equations, or applied the method of successive approximations. Of the methods taught by Aryabhata and demonstrated by his scholiast Bhaskara I, some are based on comparison of similar right-angled plane triangles and others are derived from inference. Brahmagupta is probably the earliest astronomer to have employed the theory of quadratic equations and the method of successive approximations to solving problems in spherical Astronomy.

The educational system in India at this time did not allow talented people with ability to receive training in Mathematics or Astronomy. Rather the whole educational system was family based. There were a number of families who carried the traditions of Astrology, Astronomy and Mathematics forward by educating each new generation of the family in the skills, which had been developed. It should also be noted that Astronomy and Mathematics developed on their own, separate for the development of other areas of knowledge.

Now a "Mathematical family" would have a library consisting of the writing of the previous generations. These writings would most likely be commentaries on earlier works such as the *Aryabhateeya* of Aryabhata. Many of the commentaries would be commentaries on commentaries on commentaries etc. Mathematicians often wrote commentaries on their own work. They did not aim to provide texts to be used in educating people outside the family, nor would they be looking for innovative ideas in Astronomy. Again religion was the key, for Astronomy was considered to be of divine origin and each family would remain faithful to the revelations of the subject as presented by their gods. To seek fundamental changes would be unthinkable for in asking others to accept such changes would be essentially asking them to change religious belief. Nor do these men appear to have made astronomical observations in any systematic way. Some of the texts do claim that the computed data presented in them is in better agreement with observation than that of their predecessors but, despite this, there does not seem to have been a major observational programme set up. Paramesvara in the late fourteenth Century appears to be one of the first Indian Mathematicians to make systematic observations over many years.

Mathematics however was in a different position. It was only a tool used for making Astronomical calculations. If one could produce innovative Mathematical ideas then the truths of Astronomy can be more easily exhibited. The Mathematics therefore had to lead to the same answers as had been reached before but it was certainly good if it could achieve these more easily or with greater clarity and accuracy. This meant that despite Mathematics only being used as a computational tool for Astronomy, the brilliant Indian scholars were encouraged by their culture to put their genius into advances in this topic.

A contemporary of Brahmagupta who headed the research centre at Ujjain was Bhaskara I who led the Asmaka School. This school would have the study of the works of Aryabhata as their main concern and certainly Bhaskara was one of the main commentators on the Mathematics of Aryabhata.

The 9th Century C.E., saw Mathematical progress with scholars such as Govindaswami, Mahaveera, Prthudakasvami, Sankara and Sreedhara. Some of these Mathematicians, such as Govindaswami and Sankara were commentators on the text of Bhaskara I while Mahaveera was famed for his updating of Brahmagupta's book. This period saw developments in sine tables, solving equations, algebraic notation, quadratics, indeterminate equations and improvements to the number systems. The agenda was still basically that set by Aryabhata and the topics being developed those in his work.

The main Mathematicians of the 10th Century C.E., in India were Aryabhata II and Vijayanandi, both adding to the understanding of sine tables and Trigonometry to support their astronomical calculations. In the 11th Century C.E., Sreepati and Brahmadeva were major figures but perhaps the most outstanding of all was Bhaskara II in the 12th Century C.E. He worked on Algebra, number systems and Astronomy. He wrote wonderful texts illustrated with Mathematical problems, some of which we present in his biography and he provided the best summary of the Mathematics and Astronomy of the classical period.

Bhaskara II may be considered the high point of Indian Mathematics but at one time this was all that was known: For a long time Western scholars thought that Indians had not done any original work till the time of Bhaskara II. This is far from the truth. Nor has the growth of Indian Mathematics stopped with Bhaskara II. Quite a few results of Indian Mathematicians have been rediscovered by Europeans. For instance, the development of number theory, the theory of indeterminate infinite series expressions for sine, cosine and tangent, computational Mathematics, etc.

Following Bhaskara II there was over 200 years before any other major contributions to Mathematics were made on the Indian subcontinent. In fact for a long time it was thought that Bhaskara II represented the end of Mathematical developments in the Indian subcontinent until modern times. However in the second half of the 14th Century C.E., Mahendra Soori wrote the first Indian treatise on the astrolabe and Narayana wrote an important commentary on Bhaskara II, making important contributions to algebra and magic squares. The most remarkable contribution from this period, however, was by Madhava who invented Taylor series and rigorous Mathematical analysis in some inspired contributions. Madhava was from Kerala and his work there inspired a school of followers such as Nilakantha and Jyesthadeva.

Some of the remarkable discoveries of the Kerala Mathematicians include: a formula for the ecliptic; the Newton-Gauss interpolation formula; the formula for the sum of an infinite series; Lhuilier's formula for the circum-radius of a cyclic quadrilateral. Of particular interest is the approximation to the value of π which was the first to be made using a series. Madhava's result which gave a series for π, translated into the language of modern Mathematics, reads:

$$\pi R = 4R - 4R/3 + 4R/5 - \dots$$

This formula, as well as several others referred to above, were rediscovered by European Mathematicians several centuries later. Madhava also gave other formulae for π, one of which leads to the approximation 3.14159265359.

The first person in modern times to realise that the Mathematicians of Kerala had anticipated some of the results of the Europeans on the calculus by nearly 300 years was Charles Whish in 1835. Whish's publication in the "Transactions of the Royal Asiatic Society of Great Britain and Ireland" was essentially unnoticed by historians of Mathematics. Only 100 years later in the 1940s did historians of Mathematics look in detail at the works of Kerala's Mathematicians and find that the remarkable claims made by Whish were essentially true. Indeed the Kerala Mathematicians had, as Whish wrote:-

... laid the foundation for a complete system of fluxions ...

and these works:

... abound with fluxional forms and series to be found in no work of foreign countries.

There were other major advances in Kerala at around this time. Citrabhānu was a 16th Century C.E., Mathematicians from Kerala who gave integer solutions to twenty-one types of systems of two algebraic equations. These types are all the possible pairs of equations of the following seven forms:

- $x + y = a$,
- $x - y = b$,
- $xy = c$,
- $x^2 + y^2 = d$,
- $x^2 - y^2 = e$,
- $x^3 + y^3 = f$ and
- $x^3 - y^3 = g$.

For each case, Citrabhānu gave an explanation and justification of his rule as well as an example. Some of his explanations are algebraic, while others are geometric.

Now we have presented the latter part of the history of Indian Mathematics in an unlikely way. That there would be essentially no progress between the contributions of Bhaskara II and the innovations of Madhava, who was far more innovative than any other Indian Mathematician producing a totally new perspective on Mathematics, seems unlikely. Much more likely is that we are unaware of the contributions made over this 200 year period which must have provided the foundations on which Madhava built his theories.

The world's understanding of the contributions of Indian Mathematicians has changed markedly over the last few decades. Much more work needs to be done to further our understanding of the contributions of Mathematicians whose work has sadly been lost, or perhaps even worse, been ignored. Indeed work is now being undertaken and we should soon have a better understanding of this important part of the history of Mathematics.

2.1. Oral tradition

Mathematicians of ancient and early medieval India were almost all Saṁskṛit pundits who were trained in Saṁskṛit language and literature and possessed "a common stock of knowledge in grammar (*vyākaraṇa*), exegesis (*mīmāṃsā*) and logic (*nyāya*)". Memorisation of "what is heard" (*śruti* in Saṁskṛit) through recitation played a major role in the transmission of sacred texts in ancient India. Memorization and recitation was also used to transmit philosophical and literary works, as well as treatises on ritual and grammar. Modern scholars of ancient India have noted the "truly remarkable achievements of the Indian pundits who have preserved enormously bulky texts orally for millennia".

2.2. Styles of Memorisation

Prodigious energy was expended by ancient Indian culture in ensuring that these texts were transmitted from generation to generation with inordinate fidelity. For example, memorisation of the sacred *Vedas* included up to eleven forms of recitation of the same text. The texts were subsequently 'proof-read' by comparing the different recited versions. Chanting of *Vedas* has some Mathematical sequence – like:

- *Samhitā Pāṭha* – reciting the *Vedas* in a straight sentence.
- *Pada Pāṭha* – reciting the *Veda mantras* word by word instead of joining the *padas* (words).
- *Krama Pāṭha* – reciting the *Veda mantras* – in a sequence – first and second words, second and third words, and so on. The sequence of words goes: 1-2, 2-3, 3-4, 4-5, etc.
- *Jaṭā Pāṭha* (literal meaning is "mesh recitation") – reciting the *Veda mantras* – the first two words are recited together and then the words are recited in a reverse order and then again in the original order and so on. The sequence of words goes: 1-2-2-1-1-2, 2-3-3-2-2-3, 3-4-4-3-3-4, etc.
- *Dhvaja-pāṭha* (literal meaning is "flag recitation") – a sequence of *n* words are recited (and memorized) by pairing the first two and last two words and then proceeding as: $Pada_1Pada_2$, $Pada_{n-1}Pada_n$; $Pada_2Pada_3$, $Pada_{n-3}Pada_{n-2}$; ....; $Pada_{n-1}Pada_n$, $Pada_1Pada_2$;
- *Ghana Pāṭa* (literal meaning is "dense recitation")– reciting the *Veda mantras* in the order of the sequence: 1-2-2-1-1-2-3-3-2-1-1-2-3, 2-3-3-2-2-3-4-4-3-2-2-3-4, etc. This type is supposed to be most complex form of recitation.

This clearly indicates that Mathematics is inbuilt in Vedas and they are inseparable. That these methods have been effective, is testified to by the preservation of the most ancient Indian religious text, the *Ṛgveda* (1500 B.C.E.), as a single text, without any variant readings. Similar methods were used for memorizing Mathematical texts, whose transmission remained exclusively oral until the end of the *Vedic* period (500 B.C.E.).

2.3. The *Sūtra* genre

Mathematical activity in ancient India began as a part of a "methodological reflexion" on the sacred Vedas, which took the form of works called *Vedāṇgas*, or, "Ancillaries of the *Veda*" (7th - 4th Century B.C.E.). The need to conserve the sound of sacred text by use of *śikṣā* (phonetics) and *chandas* (metrics); to conserve its meaning by use of *vyākaraṇa* (grammar) and *nirukta* (etymology); and to correctly perform the rites at the correct time by the use of *kalpa* (ritual) and *jyotiṣa* (Astronomy), gave rise to the six disciplines of the *Vedāṇgas*. Mathematics arose as a part of the last two disciplines, ritual and Astronomy (which also included astrology). Since the *Vedāṇgas* immediately preceded the use of writing in ancient India, they formed the last of the exclusively oral literature. They were expressed in a highly compressed mnemonic form, the *sūtra* (literally, 'thread'):

The knowers of the *sūtra* know it as having few phonemes, being devoid of ambiguity, containing the essence, facing everything, being without pause and unobjectionable.

Extreme brevity was achieved through multiple means, which included using ellipsis "beyond the tolerance of natural language", using technical names instead of longer descriptive names, abridging lists by only mentioning the first and last entries and using markers and variables. The *sūtras* create the impression that communication through the text was "only a part of the whole instruction. The rest of the instruction must have been transmitted by the so-called *Guru-shishya parampara*, "uninterrupted succession from teacher (*guru*) to the student (*śisya*) and it was not open to the general public" and perhaps even kept secret.

*****

3. Representation of Numerals in Saṁskṛit

यथा शिखा मयूराणां, नागानां मणयो यथा ।
तदवद् वेदाङ्गशास्त्राणाम् गणितं मूर्धनि स्थितम् ॥
वेदाङ्ग ज्योथिषम्

Yathāśikhą mayoorāṇām nāgānām maṇayo yathā |
Tadvad vedāngaśāstrāṇām gaṇitam moordhani sthitam ||
Vedānga Jyothiṣa – Lagadha, Verse 35

"Like the crest of the peacock, like the gem on the head of a snake, so is Mathematics at the head of all knowledge".

Mathematics is universally regarded as the science of all sciences and "the princess of definiteness and clarity. Everything that the greatest minds of all times have accomplished towards the comprehension of forms by means of concepts is gathered into one great science, Mathematics". J.F. Herbert.

Indian Mathematics belongs not only to a hoary antiquity but is a living discipline with a potential for manifold modern applications. Indian Mathematicians gave the world the numerals, which are now in use universally. The crowning glory of Indian Mathematics was the invention of zero and the introduction of decimal notation without which Mathematics as a scientific discipline could not have made this much headway.

"A Mathematical equation had a meaning because it expressed a thought of God" – S. Ramanujam, the man who knew infinity. In keeping with the ingrained tradition of Mathematics in India, this genius was one of the greatest Mathematicians of our time. The mystic in him blazed new Mathematical trails in Cambridge

University in the second decade of the 20th Century C.E., even though he did not himself possess a University degree.

In ancient times, throughout India, almost all the Mathematics related books were written using three types of number systems viz., *Kaṭapayādi Saṅkhyā*, *Bhūta Saṅkhyā* and *Āryabhaṭīya Saṅkhyā*.

3.1. Kaṭapayādi Saṅkhyā:

Kaṭapayādi – कटपयादि, (beginning with *Ka*, *Ṭa*, *Pa* and *Ya*), is one of the general methods of representing numerals through alphabets in Saṁskṛit. This method was pronounced by the sage *Vararuchi*. The nine numbers and the naught were represented by the letters in the Saṁskṛit consonants (as described in the table below). This enabled the ancient writers of Saṁskṛit poetry to write long numbers as words and verses made up of letters. This was done to help the learning of long lists of numbers, to convey secret messages and to narrate the numbers poetically.

Vedic knowledge is in the form of *ślokas* (verses) or poems in Saṁskṛit verses. A number was encoded using consonant groups of the Saṁskṛit alphabet and vowels were provided as additional latitude to the author in poetic composition.

Following verse found in *Śaṅkaravarman's Sadratnamāla* explains the mechanism of the system:

नञावचश्च शून्यानि संख्या: कटपयादय: |
मिश्रे तूपान्त्यहल् संख्या न च चिन्त्यो हलस्वर: ||

Nanyāvacaśca śūnyāni saṃkhyāḥ kaṭapayādayaḥ |
Miśre tūpāntyahal saṃkhyā na ca cintyo halasvaraḥ ||

Na-न, *ña*-ञ and *a*-अ-s i.e. vowels represent zero. The (nine) integers are represented by consonant group beginning with *ka, ṭa, pa, ya*. In a conjunct consonant, the last of the consonants alone will count. A consonant without vowel is to be ignored.

The coding key is also given as ***kādi nava, tādi nava, pādi paṅchaka, yādyaṣṭakaśca kṣa śūnyam*** – translated as below:

- The nine letters starting with *'ka'*
- The nine letters starting with *'ṭa'*
- The five letters starting with *'pa'*
- The eight letters starting with *'ya'*
- And the letter *'kṣa'* meaning zero.

Note: The letters *ka* to *jha* have the values from 1 to 9. The tenth letter *ṅa* has value zero. Similar is the case with *na.* That these letters have the value zero has not been explicitly stated in this verse and left to be understood. However, the conjunct consonant *kṣa* also has the value zero, which unless stated could not be inferred, due to which it has been openly mentioned.

Table 1: *Kaṭapayādi* Numerals

क *ka*	ख *kha*	ग *ga*	घ *gha*	ङ *ṅa*
1	2	3	4	5
च *ca*	छ *cha*	ज *ja*	झ *jha*	ञ *ña*
6	7	8	9	0
ट *ṭa*	ठ *ṭha*	ड *ḍa*	ढ *ḍha*	ण *ṇa*
1	2	3	4	5

त *ta*	थ *tha*	द *da*	ध *dha*	न *na*
6	7	8	9	0
प *pa*	फ *pha*	ब *ba*	भ *bha*	म *ma*
1	2	3	4	5
य *ya*	र *ra*	ल *la*	व *va*	
1	2	3	4	
श *śa*	ष *ṣa*	स *sa*	ह *Ha*	क्ष *kṣa*
5	6	7	8	0

This method operates as below:

- Numbers corresponding to each letter of any word is taken from the table
- They are written from left to right serially
- They are transposed from right to left
- Consonants have numerals assigned as per the above table. Every letter has unique number value, but the numbers can be represented by different letters. For example, *ba* (ब) is always three (3) whereas 5 can be represented by either *ṅa* (ङ) or *ṇa* (ण) or *ma* (म) or *śa* (श).
- All vowels like *a* (अ) and *ṛ* (ऋ) are assigned value zero (0).
- In case of a conjunct, consonants attached to a non–vowel will be valueless. For example, *kyā* (क्या) is formed by *k* (क्) + *ya* (य) + *a* (अ). The only consonant standing with a vowel is *ya* (य). Hence the corresponding numeral for *kyā* (क्या) will be 1.
- There is no way of representing decimal separator in this system.

- Indians used the Hindu–Arabic numeral system for numbering, traditionally written in increasing place values from left to right. This is as per the rule *aṅkānām vāmato gatiḥ* (अङ्कानाम् वामतो गति:) which means numbers go from left to right.
- By assigning more than one letter to one numeral and nullifying certain other letters as valueless, this system provides the flexibility in forming meaningful words out of numbers, which can be easily remembered.
- There is no way of representing Decimal separator in the system.

The modern day computers work on ASCII code, which translates all the characters into numbers so that they can be represented in binary system and processing can take place.

An exactly similar code has been in vogue from the letters of English alphabet also.

Table 2: English Alphabets Representing Numbers

Numbers	English Alphabets
1	b or p or ph
2	c or q or ch
3	d or r or sk
4	f or s or sh
5	g or t or th
6	h or v or gh
7	j or w or wh
8	k or x or kn
9	l or z
0	m or n or ng

The way the code works is:

- There is a choice of letters for each number as shown in the table above.
- The vowels a, e, i, o, u and the letter 'y' do not stand for any number.

Example:

fat = 45 (f = 4 and t = 5, a ignored)
father = 453
long = 90

Similar code is being used in telephones (both in desk phones and mobiles as well) nowadays as below:

Numbers	English Alphabets
2	a or b or c
3	d or e or f
4	g or h or i
5	j or k or l
6	m or n or o
7	p or q or r or s
8	t or u or v
9	w or x or y or z

Code words are used to remember the numbers to dial.

3.2. *Bhūta Saṅkhyā* (Cryptic Method)

Bhūta saṅkhyā is a method of recording numbers using ordinary words having connotations of numerical values. The method was popular among Indian Astronomers and Mathematicians since ancient times. Saṁskṛit was the language from which words were chosen to write numbers in the *Bhūta saṅkhyā* system. The system has been described as the "concrete number notation" for the representation of numbers.

The number 'two' can be associated with the word 'eye' as every human being has two eyes. Thus every Saṁskṛit word having the meaning 'eye' was used to denote 'two'. All words synonymous with 'earth' could be used to signify the number 'one' as there is only one earth. Concepts, ideas, and objects from all facets of Indian cultural experience – mythological, purānic, literary, religious, etc. – were harvested to generate number-connoting words. As an illustration, every Saṁskṛit word indicating an 'arrow' has been used to denote 'five' as *Kāmadeva* (Cupid), the Hindu deity of love, is traditionally depicted as a warrior carrying five arrows of flowers. The Saṁskṛit word *anuṣṭubh* has been used to signify 'eight' as it is the name of a meter with eight syllables in a foot. Any Saṁskṛit word for 'tooth/ teeth' could be used to denote 32 as a grown-up man has a full set of 32 teeth. Terms implying 'the gods' were used to indicate 33 as it is believed that the number of *devas* (gods) is 33 crores.

Single words indicating smaller numbers were strung together to form phrases and sentences for representing arbitrary large numbers. This formation of large numbers was accomplished by incorporating the place value system into the scheme. While decoding numbers encoded in the *Bhūta saṅkhyā* system, one should bear in mind the specialty of the Indian method of writing numbers. The various digits were written from left to right; that is, the digit with the lowest place value is written as the left most digit. The various digits of a large number are arranged from left to right in the increasing order of the place value. This specialty, succinctly indicated by the Saṁskṛit adage *aṅkūnāṃ vāmato gatiḥ*, has been extended to the *Bhūta saṅkhyā* system also.

A potential user of the system had a multitude of words to choose from for denoting the same number. The mapping from 'words' to 'numbers' is many-to-one. This has facilitated the embedding of numbers in verses in Indian treatises on Mathematics and Astronomy. This helped in memorising large tables of numbers required by Astronomers and Astrologers.

The earliest reference employing object numbers is approximately 269 C.E. Saṁskṛit text, *Yavanajātaka* (literal meaning "Greek horoscopy") of Sphujidhvaja, a versification of an earlier 150 C.E. Indian prose adaptation of a lost work of Hellenistic astrology.

In *Bhūta Saṅkhyā* the numbers are represented by Saṁskṛit words and/ or their synonyms. This is also called *Paribāṣa* Method.

For example all of us across the globe can easily relate *James Bond* to the number 007. Exactly a similar method is used in representing the numbers in Saṁskṛit through *Bhūta Saṅkhyā*.

Some of the words in Saṁskṛit correspond to a specific number for the reason stated against each in the table below.

Most of the names represent the corresponding mythological, historical, scientific and philosophical information.

Table 3: *Bhūta Saṅkhyā* Numerals

Saṁskṛit Word	Numeral Represented	Remarks
Vyoma, or *Ṣūnya*, or *Ambara*, or *Kha*	0	Null set

Saṁskṛit Word	Numeral Represented	Remarks
Pranava	1	The *Pranava Mantra 'Om'* has only one letter.
Rūpam, *Bhūmi* (Earth) or *Sasi* or *Indu* (Moon)		There is only one Earth, one Moon and one Sun
Yama	2	Doubled *Varga* letter when followed by a fifth letter of *Varga*
Nayanam (eyes), *Hastam* or *Karnam* (hands), or *Pādam* (Legs)		Organs of the body which are in duplicate
Yugmam		Couple
Guṇa	3	*Sathwa, Rajas, Tamas*
Mūrthy		*Brahma, Viṣnu, Siva*
Rāma		*Balarāma, Śrirāma, Parasurāma*
Agni		*Āhavanīyāgni, Gārhapatyāgni, Dākṣināgni*
Vyāhriti		*Bhūḥ, Bhuvaḥ, Suvaḥ*
Veda	4	*Ṛg, Yajus, Sāma, Atharva*
Samudra		*Lavaṇa, Kṣīra, Ikṣu, Dadhi*
Ishu or *Śara*	5	*Ishu* or *Śara* is the name of the 5 arrows held by *Manmatha* (Cupid) made of 5 different

Saṁskṛit Word	Numeral Represented	Remarks
		flowers viz. *Aravindam*, *Aśokam*, *Cūtam* (high breed Mango), *Navamallika* (Jasmine) and *Nīlotpalam*
Bhūta		*Panca Bhūtas* – Earth, Air, Water, Ether and Fire
Anga		*Panca Angas – Thithi, Vāra, Nakṣatra, Yoga,* and *Karaṇa*
Prāna		*Prāna, Apāna, Vyāna, Udāna* and *Samāna*
Indriya		*Panca Indriyas – Jñānendriyas* and *Karmendriyas* (Five sense organs and five organs of action)
Āranya		*Badrikāraṇyam* (*Ilandai*), *Vamsāraṇyam* (Bamboo), *Vedāraṇyam* (*Ālangādu*), *Vīkṣāraṇyam* (*Īkkādu*) and *Naimisāraṇyam* (*Darbai*)
Sarpa	6	6 important snakes – *Ananta*, *Takṣaka*, *Kārkotaka*, *Vāsuki*, *Ādisesha* and *Piṅgala.*
Rithu,		*Vasanta*, *Grīṣma*, *Varṣā*, *Śarad*, *Hemanta* and *Śiśira.*

Saṁskṛit Word	Numeral Represented	Remarks
Rasa,		*Madhura, Āmla, Lavaṇa, Kaṭu, Kaṣāya* and *Tikta*
Chakra		*Mūladhara, Swādhiṣtāna, Maṇipūraka, Anāhatha, Viśuddhi* and *Ājya,*
Rṣi or *Muni*	7	*Sapta Rṣi* – *Kāshyap, Atri, Baradwāja, Vishwāmitra, Goutama, Jamadagni* and *Vashishta* (the names of *Sapta Rṣis* are mentioned differently by different schools – these are the commonly accepted names)
Adri or *Giri*		*Sapta Parvatas*– *Vṛṣabādri, Nīlādri, Anjanādri, Śeṣādri, Garudādri, Nārāyanādri, Venkatādri*
Svara		*Sapta Svaras*
Vāsara		Seven days of a week
Lakshmee	8	Eight *Lakshmees*
Dik		Eight directions
Vasu		*Aṣta Vasus* (*Agni, Prithivi, Vāyu, Antarikṣa, Āditya, Dyou, Candramā* and *Nakṣatras*)
Gaja		*Aṣta Dik Gajas*
Aśwa		*Aṣta Aśwas*

Saṁskṛit Word	Numeral Represented	Remarks
Graha *Randhra* *Nanda*	9	*Navagrahas* *Navadvāras* Nine Nandas of Nanda Dynasty
Avatār	10	*Daśāvatāras* of Lord Viṣṇu
Rudra or *Mahādeva* or *Śaṅkara*	11	*Ekādaśa Rudras – Mahādeva, Śiva, Śrīrudra, Śaṅkara, Nīlalohita, Īśāna, Vijaya, Bhīma, Devadeva, Bhavodhbhava, Ādityātmika*
Āditya or *Sūrya* or *Ravi*	12	12 *Ādityas* corresponding to 12 *Rāśis* (*Mitra, Ravi, Sūrya, Bhānu, Khaga, Puṣṇa, Hiranyagarbha, Marīci, Āditya, Savitr, Arka and Bhāskara*)
Viṣvedeva	13	13 *Viṣvedevas*
Manu	14	14 *Manus*
Thithi	15	15 *Thithis* from *Prathamā* to *Amavāsyā* or *Pūrnimā*
Jina	24	24 *Tirthaṅkaras*
Bhā, Nakṣatra	27	27 stars – *Aśvinī* to *Revatī*
Danta	32	32 teeth
Samvatsaram or Varṣam	60	60 years
Kalā	64	64 arts

Examples:

The numbers are written in the order stated and then transposed. In other words the digits in the unit, 10th, 100th, 1000th, 10000th places are written in the same order of the Saṁskṛit words.

Example 1: *Sara* (5) *adri* (7) *rāma* (3) *anala* (3) *yojanāni kuvayu kakṣya parita*: =
the circumference of the earth is 3375 (transpose of 5733) *yojanās*.
Example 2: The number of days in a *mahāyuga*
= *vyoma– śūnya– śara–adri–indu–randhra–adri–adri–śara–indavaḥ*
= *vyoma*(0) *śūnya* (0) *śara* (5) *adri* (7) *indu* (1) *randhra* (9) *adri* (7) *adri* (7) *śara* (5) *indava* (1)
= 1577917500 (transpose of 0057197751) days.
Example 3: *kha rasa svara muni bhūta rūpa indu manavaḥ*
= 141157760;
Example 4: *rūpa nanda agni vasu rudra hate*
= multiply with 118391 (*hate* means to multiply)
Example 5: *sara rasa nava bhū bha ambara karṇa nadi Veda hṛte* =
divide by 4420271965 (*hṛte* means to divide)
Example 6: Let us consider a certain number used extensively in Indian Astronomy. Varāhamihira (505 – 587 C.E.), an Indian Astronomer, Mathematician and Astrologer, encoded this number in *Bhūta saṅkhyā* as *kha-kh-āṣṭi-yamāḥ*.

The individual words in this are "*kha*", "*kha*", "*aṣṭi*" and "*yamāḥ*" and they denote the numbers "0", "0", "16" and "2" respectively in that order.

To obtain the modern equivalent of the number indicated by *kha-kh-āṣṭi-yamāḥ*, the four numbers have to arranged in the reverse order, namely, in the order "2", "16","0" and "0". Placing these four numbers side by side we get the number 21600. Incidentally, the number 21600 is the number of minutes in a full circle.

Example 7: The system has also been used extensively in epigraphical inscriptions in the Indian subcontinent for inscribing dates and years. As an example, in an inscription from Kalna, the date is given in *Bhūta saṅkhyā* system as *bāṇa-vyoma-dharādhar-indu-gaṇite śāke*, which means "In the Śāka year enumerated by the arrows (5), the sky (00, the mountains (7) and the moon (1)", that is, in Śāka 1705 = C.E. 1783".

The synonyms are commonly used for all the above Saṁskṛit words and the data presented in a poetical verse keeping the number of letters in each line to make the *chandas*/ rhythm/ prosody.

More than a thousand Mathematical/ Astronomical books like *Laghubhāskarīyam*, *Mahābhāskarīyam*, *Vaṭeśvara Siddhāntha*, *Āryabhaṭīyam*, *Śiṣyadhī–vṛuddhi–tantra*, etc., most commonly, use the *Bhūta Saṅkhyā* for representing the numbers. Most of the *āyurvedic*/ metallurgical books also use this method of representing numbers.

Right now there are only very few proponents of this method. Without the knowledge of this method, one cannot learn or understand the Indian Scientific heritage.

Hundreds of Mathematical theorems in use today, were already available in ancient India. But they could not be credited or copyrighted, since the numbers could not be properly decrypted. For example:

In 1671 Gregory James rediscovered the theorem represented by the arc–tangent series, which in 14th Century C.E., itself an Indian Mathematician *Mādhava* of *Sangamagrāma* had originally discovered:

$$\theta = \tan\theta - (1/3)\tan\theta + (1/5)\tan\theta - \ldots,$$

for θ lying between $-\pi/4$ and $+\pi/4$.

This formula was discovered and used by *Sangamagrāma Mādhavācārya* to calculate digits of π more than 3 centuries ahead, by using the Saṁskṛit *Sūtra*: "*Vyāsāt vana sanguṇitāt*"

This means that the diameter has to be multiplied with *vana* (= 4). Due to lack of understanding/ interpretation of the *Sūtra*, the theorem could not be comprehended.

Note: For any given circle circumference = π x d (diameter).

3.3. Āryabhaṭīya Saṅkhyā

Āryabhaṭā created his own way of representing the numbers through Saṁskṛit alphabet by about 499 C.E., in his book called Āryabhaṭīyam. Hence this system is also known as Āryabhaṭa numeration. This book has 4 *kāṇḍā*–s. This numerical representation is described in the first *kāṇḍā* called *Gītikā kāṇḍā* in 10 verses. The number corresponding to each of the Saṁskṛit alphabet is listed in the table given in Annexure 2.

The Āryabhaṭa numeration is a system of numerals based on Saṁskṛit phonemes. It was introduced in the early 6th century C.E., by Āryabhaṭa, in the first chapter titled *Gītika Padam* of his *Āryabhateeya*. It attributes a numerical value to each syllable of the form consonant + vowel possible in Saṁskṛit phonology, from *ka* = 1 up to *hau* = 10^{18}.

The basis of this number system is mentioned in the second stanza of the first chapter of *Āryabhateeya*. The *Varga* (Group/ Class) letters *Ka* to *Ma* are to be placed in the *varga* (square) places (1st, 100th, 10000th ...etc., places) and *Avarga* letters like *Ya, Ra, La..* have to be placed in *Avarga* places (10th, 1000th,...etc., places).

The *Varga* letters *Ka* to *Ma* have value from 1, 2, 3 .. up to 25 and *Avarga* letters *Ya* to *Ha* have value 30, 40, 50.. up to 100. In the *Varga* and *Avarga* letters, beyond the ninth vowel (place), new symbols can be used.

The values for vowels are as follows:

i = 100 ; *u* = 10000, *ru* = 1000000 and so on.

Āryabhata used this number system for representing both small and large numbers in his Mathematical and Astronomical calculations. This system can even be used to represent fractions and mixed fractions. For example *nga* is 1/5, *nja* is 1/10 and *Jhardam* (*jha* = 9; its half) = 4½.

Example: The number 29,97,92,458 (Twenty–nine Crores ninety–seven lakhs ninety–two thousand, four hundred fifty–eight) is represented in *Āryabhaṭīya Saṅkhyā* as:

जल घिनि झुशु झृसृ खॢ *ja–la ghi–ni jhu–śu jhṛ–sṛ khḷ*

Table 4: *Āryabhaṭīya Saṅkhyā* – an Example

ज	ल	घि	नि	झु	शु	झृ	सृ	खॢ
ja	*la*	*ghi*	*ni*	*jhu*	*Śu*	*jhṛ*	*sṛ*	*khḷ*
8	5	4	2	9	7	9	9	2
10^0	10^1	10^2	10^3	10^4	10^5	10^6	10^7	10^8
Adding the products of rows 3 & 4 above we get – 299,792,458								

The main advantage of this system is that very huge numbers can be represented using minimum number of alphabet.

Āryabhaṭa used this number system in his books on Astronomy and Astrological calculations.

This type of *Āryabhaṭīya* numbering system is not much in vogue, because of its known disadvantages like difficulty in pronunciation (unlike *Bhūta Saṅkhyā*, this is not rhyming/ poetic), meaninglessness, difficulty in remembering, etc. Hence, the findings of Āryabhaṭa in his books are not much popular, in–spite of its advantages and being innovative and scientific.

*****

4. Applications of Saṅkhyās

The *Kaṭapayādi Saṅkhyā*, *Bhūta Saṅkhyā* and *Āryabhaṭīya Saṅkhyā*-s were extensively used in Saṁskṛit books, verses, etc. This served the dual purpose – to compose the poems in the corresponding meters (*chandas*) and to denote the mathematical numbers and operations in the form of letters and words. It can be imagined how much skill is needed to write such poems. Equally knowledge in both Saṁskṛit and Mathematics is needed to decode the same. This is very much useful in cryptic or cypher communication.

Here are some of the applications of *Saṅkhyā*-s:

- Important dates were remembered by converting them using *Kaṭapayādi* system. These dates are generally represented as number of days since the start of *Kali Yuga*. It is sometimes called *kalidina Saṅkhyā*.
- The Malayalam calendar known as *kollavarsham* was adopted in Kerala beginning from 825 C.E., revamping some existing calendars. This date is remembered as *āchārya vāgbhadā*, converted using *Kaṭapayādi* into 1434160 days since the start of Kali Yuga.
- The oldest available evidence of the use of Kaṭapayādi System is from Grahacāraṇibandhana by Haridatta in 683 C.E.
- In some astronomical texts popular in Kerala planetary positions were encoded in Kaṭapayādi system.
- On Saṁskṛit astrolabe discovered in North India, the degrees of the altitude are marked in the Kaṭapayādi system.
- Mādhava's Sine table constructed by 14th Century C.E., by Kerala Mathematician-Astronomer Mādhava of Sangamagrama, employs the Kaṭapayādi system to enlist the trigonometric sines of various angles.

4.1 Carnatic Music:

An interesting feature of the *Melakarta Rāgas* of *Carnatic* (classical) music uses Kaṭapayādi Saṅkhyā extensively. The *Melakarta Rāga* is named so that the first two syllables of the name will give its number using Kaṭapayādi Saṅkhyā. The table in Annexure 2 lists down all the 72 *Melakarta Rāgas,* in that order.

The famous verse found in Sadratnamala describes this process - The numbers corresponding to first two letters of the *Rāga* are to be read from the Kaṭapayādi table and transposed. The number of the *Rāga* is obtained.

For example:

a. 15th *Melakarta Rāga* मायामालवगौल *Māyāmalavagoula* – The first two letters are मा (*Mā*) and या (*Yā*). The numbers corresponding to these letters are 5 and 1 = 51. By transposing 51 we get 15, which is the number for the *Rāga*.
b. 30th *Melakarta Rāga* नागनन्थिनि *Nāganandhini* – The first two letters are ना *Nā* and ग *Ga*. The numbers corresponding to these letters are 0 and 3 = 03. By transposing 03 we get 30, which is the number for the *Rāga*.

Note: In Saṁskṛit the half letters are not taken into consideration for counting/ calculations.

These *Melakarta Rāgas* are also called as *Sampoorna Rāgas*, because all the 7 *swaras* are used in both *Ārohana* and *avarohana*. Kaṭapayādi Saṅkhyā can also be used to find the *swaras* of *Ārohana* and *avarohana*.

Out of the *Swaras* ***'Sa'*** (*Shadja*) and ***'Pa'*** (*Panchama*) are fixed and here is how to get the other *swaras* from the melakarta number.

1. *Melakartas* 1 through 36 have ***Ma1*** (*Shuddha Madhyama*) and those from 37 through 72 have ***Ma2*** (*Prati Madhyama*).
2. The other notes are derived by noting the (integral part of the) quotient and remainder when one less than the *melakarta* number is divided by 6.
3. ***'Ri'*** (*Rishabha*) and ***'Ga'*** (*Gāndhāra*) positions – the *rāga* will have:
   - ***Ri1*** (*Shuddha Rishabha*) and ***Ga1*** (*Shuddha Gāndhāra*), if the quotient is 0
   - ***Ri1*** (*Shuddha Rishabha*) and ***Ga2*** (*Sādhārana Gāndhāra*), if the quotient is 1
   - ***Ri1*** (*Shuddha Rishabha*) and ***Ga3*** (*Anthara Gāndhāra*), if the quotient is 2
   - ***Ri2*** (*Chathusruthi Rishabha*) and ***Ga2*** (*Sādhārana Gāndhāra*), if the quotient is 3
   - ***Ri2*** (*Chathusruthi Rishabha*) and ***Ga3*** (*Anthara Gāndhāra*), if the quotient is 4
   - ***Ri3*** (*Shatusruthi Rishabha*) and ***Ga3*** (*Anthara Gāndhāra*), if the quotient is 5
4. ***'Da'*** (*Dhaivatha*) and ***'Ni'*** (*Shuddha Nishadha*) positions: the rāga will have:
   - ***Da1*** (*Shuddha Dhaivatha*) and ***Ni1*** (*Shuddha Nishadha*), if the remainder is 0
   - ***Da1*** (*Shuddha Dhaivatha*) and ***Ni2*** (*Kaishiki Nishadha*), if the remainder is 1
   - ***Da1*** (*Shuddha Dhaivatha*) and ***Ni3*** (*Kākali Nishadha*), if the remainder is 2
   - ***Da2*** (*Chathusruthi Dhaivatha*) and ***Ni2*** (*Kaishiki Nishadha*), if the remainder is 3
   - ***Da2*** (*Chathusruthi Dhaivatha*) and ***Ni3*** (*Kākali Nishadha*), if the remainder is 4
   - ***Da3*** (*Shatsruthi Dhaivatha*) and ***Ni3*** (*Kākali Nishadha*), if the remainder is 5

Examples

- ***Rāga Dheerasankarabharanam***

The Kaṭapayādi scheme associates *dha* = 9 and ra = 2, hence the raga's *melakarta* number is 29 (92 reversed). Now 29 < 36, hence *Dheerasankarabharanam* has ***Ma1***. Divide 28 (1 less than 29) by 6, the quotient is 4 and the remainder is 4. Hence, this *rāga* has ***Ri2***, ***Ga3*** and ***Da2***, ***Ni3***. Therefore, this rāga's scale is *Sa Ri2 Ga3 Ma1 Pa Da2 Ni3 SA*.

- **Rāga Mechakalyāni**

From the scheme *Ma* = 5, *Cha* = 6. Hence the *rāga's melakarta* number is 65 (56 reversed). 65 > 36. So *Mechakalyani* has *Ma2*. Since the rāga's number is greater than 36 subtract 36 from it.
i.e. 65 -36 = 29.
Divide 28 (1 less than 29) by 6: quotient = 4, remainder = 4.
Hence *Mechakalyāni* has the notes *Sa Ri2 Ga3 Ma2 Pa Da2 Ni3 SA*.

- **Exception for Simhendramadhyamam**

It can be seen that, as per the above calculation we should get *Sa* = 7, *Ha* = 8 giving the number 87 instead of 57 for *Simhendramadhyamam*. This should be ideally *Sa* = 7, *Ma* = 5 giving the number 57. So it is believed that the name should be written as ***Sihmendramadhyamam*** (As in the case of Bra**hm**ana in Saṁskṛit).

Figure 1: *Melakarta Rāgas* and *Kaṭapayādi*

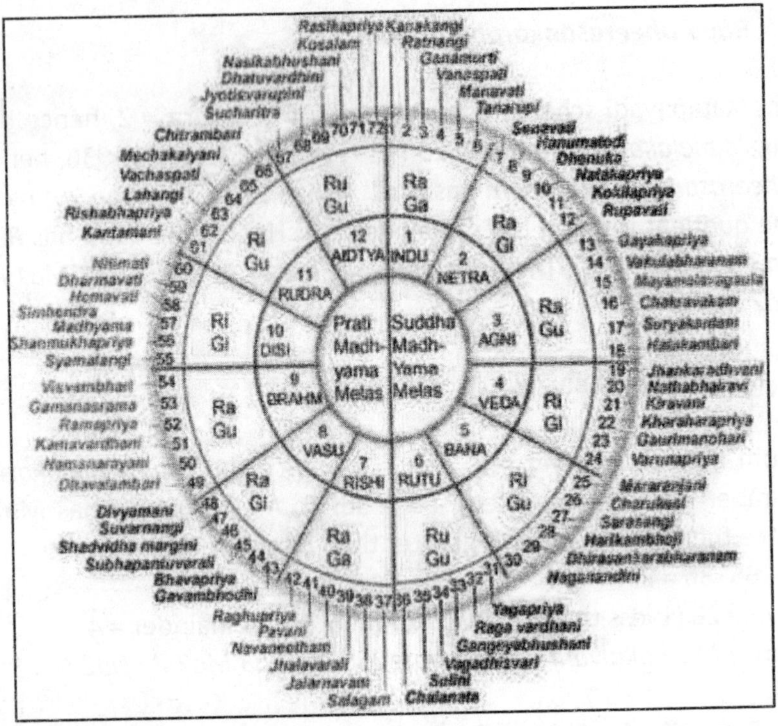

4.2 *Lalitā Sahasranāma*

It is a known fact that *Lalitā Sahasranāma Stotra* bows to goddess *Lalitā* through 1000 names. This was elucidated by Lord *Hayagrīva* himself to sage *Agastya*.

The 1000 names are contained in 182.5 verses of 2 lines each. Identifying the individual names by splitting the verses is not an easy task. The names need to be identified satisfying three conditions viz.:

- There should be exactly 1000 names not one more nor one less
- There should be no duplication

- There should be no fillers like *ca, tu, hi, eva*, etc.

To help splitting the names, there were 74 verses called *Śalākṣara–sūtra–s*. These verses used a blend of *Kaṭapayādi* and *Bhūta Saṅkhyās*. Unfortunately, these verses barring 3 or 4 are not available now.

Śrī Bhāskararāya was the first to write a *Bāṣyam* (commentaries) for *Lalitā Sahasranāma*. His *Guru, Śrī* Nṛsimhanandanātha has written 40 verses called *Paribhāṣā Maṇḍalam* in the place of *Śalākṣara–sūtra–s*. The surface meaning of each of the verse in the *Paribhāṣā Maṇḍalam* seems to be praising the glory of *Śrī Devee*. But when read in depth with a different perspective, the meaning indicates the beginning letters of the names and the number of letters in each name as well. Again the beginning letters of each of the hundredth name (1st, 101st, 201st, ...) has also been mentioned in one the verses. This clearly explains how the number system has been mapped to the Saṁskṛit letters and extensively used in verses.

For instance the words *Tripurām* and *Kulanidhim* in the first verse:

Tripurām – *pu* = 1 and *rā* = 2 – when transposed we get 21 – that is the number of names beginning with the letters *tri* is 21.

Kulanidhim – *nidhi* is *navanidhi* – that is, it indicates the number 9 – the number of names beginning with the letters *kula* is 9.

However, since they are in cryptic form, it is very difficult to interpret and understand and hence its usage is not so popular[3].

---

[3] This author has written a separate book "Paribhāsha Stora-s", decrypting all the 40 verses into 1000 names.

4.3 *Purānā-s*

The prayer verse of 18 *purānā*-s uses the Kaṭapayādi Saṅkhyā.

Nārāyanam namaskrutya, naram chaiva narottamam |
Deveem saraswateem vyāsam, tato jayamudeerayet ||

The meaning of this verse is:

Having bowed down to Lord *Shreeman Narāyana* and *Nara*, the most exalted male being (*Shree Krishna*), to the Goddess Saraswatee, and to Veda Vyās, must the word Jaya (victory) be uttered.

This is also the first verse of *Mahābhāratha*. It was originally called as *Jaya-itihāsam* and Maharishi Vyāsa himself named it Mahābhārath.

Let us focus on the highlighted word - *jaya*. The above translation says it is victory. But, as per Kaṭapayādi Saṅkhyā, *ja* = 8 and *ya* = one. i.e. 81, after transposing it is = 18.

What does this number signify? These are the *parvas* in Mahābhārata. Hence the meaning must read after prostrating to Lord Shreeman Narāyana, Shree Krishna Paramatma, the best among men, Goddess Saraswatee and Vyāsa, the narration of Mahābhārata of eighteen chapters should be read.

Even *Shreemad Bhagāvatam* has eighteen cantos and hence the same verse can be used to start Bhagāvata Parāyana also.

Shreemad Bhagāvad Geeta also has 18 chapters. The Mahābhārata battle was fought for 18 days. Hence the number 18 has lot of references in Purāna-s and the word *jaya* is applicable everywhere.

If जय *jaya* is reversed, it becomes यज *yaja*. *Yaja* means *yagna* or to perform *yagna*. Hence, anything related to *yagna* becomes *yaja* or *jaya*. If a *Kāvya* specifies *yagna* related *dharma* and *mantras* that are prescribed in *Vedas*, it can be called as *yaja*. Anything through, which a person can achieve चतुर्विध पुरुश *chaturavidha purusha* can be termed as *jayam* or *jaya*. A person is born to achieve four goals चतुर्विध पुरुश and that accomplishment is called *jayam*. These four goals are धर्म, अर्थ, काम and मोक्ष (*Dharma, Artha, Kāma* and *Moksha*). Hence, those books, which define these four goals and related *dharmas* to achieve them, are called *jayam*. Alternately, *jaya* is defined in *Anga shāstram*.

4.4 *Bāla Mantra*

Many *Tantras* are unified in declaring that the *mantra* of *Bāla* is of three letters. (*Aim kleem souh*). The eighteen *siddhas* of the yore also practised the *Bāla mantra* in their own way by chanting *Aim kleem souh* and attained the eight *siddhis*. Their practical verbatim in Tamil language runs "there is no more nourishing food other than milk; there is no more deity above *Bāla*".

By tradition the *Bāla mantra* is initiated as a six letters. This is derived by adding the reverse order of the above three letters in the end of the aforesaid mantra (*Aim kleem souh souh kleem aim*). This *mantra* is devoid of any curse of lock on the powers. Later when the *upāsana* progresses the *Shreevidyā Pancadashee mantra* is imparted to the aspirant. Then this Bāla *mantra* takes the form of the *Yoga Bāla mantra* – the *Anga mantra* (primary attendant) of the *Pancadashee mantra*. There the *mantra* has a nine letter form, which is derived by adding the three letters of the first said Bāla, which is in the normal order, after the above six letter *mantra*. (*Aim kleem souh souh kleem aim Aim kleem souh*).

According to *Kaṭapayādi Saṅkhyā*, *ba* = three and *la* is also = three. The first inference is identification of the number of *beejāskharas* in the *Bāla mantra* namely three. Since there are two three's for *ba* and *la*, adding these we find number of *beejāskhara*-s in the *Bāla mantra* (six) imparted by tradition. Multiplying them we find the number of *beejāskhara*-s (nine) in the *Anga Bāla* (*Yoga Bala*) of *Panchadashee*.

Subtracting them we get zero, which in Saṁskṛit is denoted as '*poojya*', which can also be translated as 'worthy of worship'. Hence inferred that *Bāla* is worthy of worship. Dividing them we get one, which denotes *ekam sat viprah bahuta vadanti* – the *Brahmam*, which pervades this entire universe.

Placing both the three's side by side we get thirty three, the number of divine beings in the *Devaloka*, The eleven Rudras, twelve *Ādityas*, eight *vasus* and *Ashvini* duo make up the divine pantheon (11 + 12 + 8 + 2 = 33).

Multiplying both three's we get nine, the number of planets in the skies as per tradition, Hence *Bāla* is *Graharoopinee*.

In the above calculation multiplying three thrice we find twenty seven the number of stars (*nakshatras*), hence *Bāla* is Nakshatra *roopinee*.

Adding the two threes we get six the number of *yoginidevee's* (*Hākinee,* etc.) in the six body *chakras*, therefore Bāla is Yoginee roopinee.

The number three denotes the main three *peetas* (seats of divine powers) *Kāmagiri, Poornagiri* and *Jalandhara*. Hence *Bāla* is *Peeta roopinee*.

4.5 *Nāma Shāstra* (naming conventions)

The *Kaṭapayādi* system is used in naming newborns called as *nāma nakshatra*. From the very name, the birth star of the baby can be found out. It has some more significance with the horoscope.

The name is the first *mantra*. *Nāmashāstra* is therefore *Mantra shāstra*. The name has its basis in 2, 3 or 4 syllables as being movable, dual or fixed signs of Astrology. This sequence is repeated for names which are longer (or shorter, viz. 1 = fixed sign). Only the FIRST name is considered here and family names and titles are omitted. However, if there are TWO first names, then both are counted.

Name is the first tangible result, which the eternal soul attains with human birth in the world of *Māyā*. This gives the identity and individuality. Whenever anyone calls with that name, it fills up the mind with some very unique perceptions regarding the self. The perceptions are very subjective and very unique and the persons near us interact with us in a very specific way. This is a common occurrence that, a person might have more than one name in the home, among friends and in his work circle. This is seen that the way the people or the world behaves considerably change with the name.

Let us analyse the *Kaṭapayādi* table:

Kaṭapayādi Varga										
Value	0	0	0	0	0	0	0	0		
Vowels	Overlord-Sun									
	अ	आ	इ	ई	उ	ऊ	ऋ	ॠ		
	a	*ā*	*i*	*Ee*	*u*	*oo*	*ru*	*roo*		
	ऌ	ॡ	ऐ	ओ	ओ	औ	ऑ	अः		
	~lu	*~lU*	*ai*	*O*	*O*	*au*	*~o*	*aḥ*		
	The vowels have no body and indicate the soul, therefore their number is 0. Zero is the great void or *mahāshoonya* representing the absolute.									
Values	1	2	3	4	5	6	7	8	9	0
Consonants - have body	Overlord-Mars					Overlord-Venus				
	क	ख	ग	घ	ङ	च	छ	ज	झ	ञ
	ka	*kha*	*ga*	*gha*	*~g*	*ca*	*cha*	*ja*	*jha*	*~ja*
	Overlord-Mercury					Overlord-Jupiter				
	ट	ठ	ड	ढ	ण	त	थ	द	ध	न
	ṭ	*ṭha*	*ḍa*	*ḍha*	*ṇa*	*ta*	*tha*	*da*	*dha*	*Na*
	Overlord-Saturn									
	प	फ	ब	भ	म					
	pa	*pha*	*ba*	*bha*	*ma*					
	Overlord-Moon									
	य	र	ल	व	श	ष	स	ह	क्ष	
	ya	*ra*	*la*	*va*	*sha*	*Sha*	*sa*	*ha*	*kSha*	

To find the name sign (*nāmarāshi*):

- Write the name in *devanāgari* script and remove the half consonants. For instance in the name *Vishnu*, the letter *sh* is half. Only consonants followed by vowels have life (*svara*), the remaining must be excluded. Therefore *va* and *na* must be considered for calculation.

- Vowels initiating a name get a value of 0, the rest are omitted. For instance in the name *Anna,* the letter *a* gets the value of 0 whilst *na* gets 5.
- The numbers once attained must always be reversed, For instance in the name *Anna* the numbers were 0 & 5, which when reversed gives 50.
- To find the sign/ rāshi remove multiples of 12 from the number and keep only the remainder. The remainder will be the sign counted from Aries. If the reminder obtained is "0", then 12 becomes the reminder. The remainder when counted from Aries gives the sign on which the name has predominant influence.
- In the above example 50 divided by 12 gives the remainder 2 and hence the sign is Taurus.
- The *nāmarāshi* will show the impact of the name on life of the person. The house, planets, etc., in that sign or lording it will raise due to the name.
- Picking a good name is important.

Further divide the reversed number by 9. The reminder so obtained shall show the predominating planet influencing the name.

1- Sun, 2- Moon, 3- Jupiter, 4- Rahu, 5- Mercury, 6- Venus, 7- Ketu, 8- Saturn and 9- Mars.

In the above example 50 divided by 9 gives the remainder 5 and hence the ruling planet is Mercury.

Let everyone choose a good name and get all the benefits.

*****

5. Vedic Mathematics

The *Vedas* are the most ancient record of human experience and knowledge, passed down orally for generations. It is believed that this should have been written thousands of years ago. Medicine, Architecture, Astronomy and many other branches of knowledge, including Mathematics, are dealt with in these texts. It is not surprising that the country credited with introducing the current number system and the invention of perhaps the most important Mathematical symbol '0', may have more to offer in the field of Mathematics.

Figure 2: HH Jagadguru Swāmi Śrī Bhārati Kṛṣṇa Tīrthāji Mahārāj

Vedic Mathematics, compiled by, **HH Jagadguru Swāmi Śrī Bhārati Kṛṣṇa Tīrthāji Mahāraj (जगद्गुरु स्वामि श्री भारती कृष्ण तीर्थाजी महराज:** March, 1884 – February 2, 1960) of *Govardhan Pīṭha* is a monumental work. The phrase "**compiled by**" and not "written by" is consciously used, for the obvious reason that this was available in *Vedas* and *Śaṅkarācārya* discovered the formulae and did not invent them.

In his deep–layer explorations of cryptic *Vedic* mysteries relating especially to the calculus of shorthand formulae and their neat and ready application to practical problems, the *Śaṅkarācārya* shows the rare combination of the probing insight of revealing intuition of a *Yogi* with the analytic acumen and synthetic talent of a Mathematician.

With the *Śaṅkarācārya,* we belong to a race, now fast becoming extinct, of diehard believers who think that the *Vedas* represent an inexhaustible mine of profoundest wisdom in matters, which are both spiritual and temporal; and that this store of wisdom was not, as regards its assets of fundamental validity and value at least, gathered by the laborious inductive and deductive methods of ordinary systemic enquiry, but was direct gift of revelation to seers and sages who in their higher reaches of *Yogic* realisation were competent to receive it from a source, perfect and immaculate.

Vedic Mathematics offers a fresh and highly efficient approach to Mathematics covering a wide range – right from elementary multiplication till relatively advanced topics like the solution of non–linear partial differential equations. But the *Vedic* scheme is not simply a collection of rapid methods; it is a system, a unified approach. The essence of the system is that it is based on sixteen *sūtra*–s – a *sūtra* is a terse statement of an important point or principle.

Whether or not the *Vedas* are believed as repositories of perfect wisdom, it is unquestionable that the *Vedic* race lived not merely as a pastoral folk possessing a partially developed culture and civilisation. The *Vedic* seers proved themselves adepts in all levels and branches of knowledge, both theoretical and practical. For example, they had their varied objective as science both pure and applied.

A concrete illustration is considered. Suppose in a time of drought we require rains by artificial means. The modern scientists have their own theory and art (technique) for producing the result. The old seer scientists also had these both, but different from the form now available. They had science and technique, called *Yajna*, in which *Mantra, Yantra* and other factors must co–operate with Mathematical determinates and precision. For this purpose, they had developed the six auxiliaries of the *Vedas* in each of which the Mathematical skill and adroitness, occult or otherwise, play a decisive role. The *sūtra*–s lay down the shortest and surest lines. The correct intonation of the *Mantra*, the correct configuration of the *Yantra* (in the making of the *Veda*, etc., e.g. the quadrate of a circle), the correct time or astral conjunction factor, the correct rhythms, etc., are all important aspects to be noted. All had to be perfected so as to produce the desired results effectively and adequately. Each of these required precise Mathematical calculations. The modern technician has his logarithmic tables and mechanic's manuals, the old *Yajnik* had their *Sūtra*–s.

The revered *Śaṅkarācārya* used to say that he had reconstructed the sixteen Mathematical formulae from the *Atharva Veda* after assiduous research and *Tapas* (austerity/ penance) for about eight years in the forests surrounding Śringeri. Obviously these formulae are not to be found in the present editions of *Atharva Veda*. They were actually reconstructed, on the basis of intuitive revelation, from materials scattered here and there in the *Atharva Veda*. According to him these aphorisms are contained in the *Parisiṣṭa* (the appendix portions) of the *Atharva Veda*.

Swāmiji denoted the value of π (*pi*) as:

गोपीभाग्य मधुव्रात श्रृङ्गिशो दधिसन्धिग ।

खलजीवति खाताव गलहाल रसन्ध्र ॥

Gopībhāgya madhuvrāta sṛngīśo dadhisandhiga |
Khalajīvati khātāva galahāla rasandhra || *Bhāratī Tīrta*

The apparent translation is as follows: "The fortune of the *Gopikas*, the destroyer of the demon *Madhu*, protector of cattle, the one who ventured into the ocean depths, destroyer of evil does, one with plough on the shoulder and the bearer of nectar, May you protect us". Again the same verse, using *Kaṭapayādi* method (the method is explained in the previous chapter), gives the value of π (*pi*) upto 31 decimal places:

go = 3, *pi* = 1, *bhā* = 4, *ya* = 1, *ma* = 5, *dhuv* = 9 and so on

π (*pi*) = 3.1415926535897932384626433832792

5.1. Salient Features of Vedic Mathematics

Vedic Mathematics is becoming increasingly popular as more and more people are introduced to the beautifully unified and easy calculation methods. Some interesting salient features in the very words of HH Jagadguru Śri Bhārati Kṛṣṇa Tīrthaji Mahāraj from the preface of his book on "*Vedic* Mathematics" are worth mentioning:

"*Vedas*, the oldest 'Religious' scriptures of the whole world, claim to deal with all branches of learning whether spritual or temporal and to give the earnest seeker after knowledge all the requisite instructions and guidance in full detail and on scientifically – nay – Mathematically – accurate lines in them all and so on".

The very word *Veda* has the derivational meaning, i.e. the fountain–head and illimitable store–house of all knowledge. This derivation, in effect, means, connotes and implies that the *Vedas* should contain within themselves all knowledge needed by mankind. That is, it connotes and implies that our ancient Indian *Vedic* lore should be all round complete, perfect and be able to throw light on all matters which any aspiring seeker after knowledge can possibly seek to be enlightened on.

1. The *Sūtra*–s (aphorisms) apply to almost every branch of Mathematics.
2. The *Sūtra*–s are easy to understand, easy to apply and easy to remember and the whole work can be truthfully summarised in one word as 'Mental'.
3. Even as regards complex problems involving a good number of Mathematical operations (consecutively or even simultaneously to be performed), the time taken by the *Vedic* method will be a third, a fourth, a tenth, or even a much smaller fraction of the time required according to modern (i.e. current) methods.
4. And in some very important and striking cases, sums requiring 30, 50, 100 or even more numerous and cumbrous 'steps' of working (according to the current methods) can be answered in a single and simple step of work by the *Vedic* method. Children (of only 10 or 12 years of age) merely look at the sums written on the blackboard and immediately shout out and convey the answers. This is because, as a matter of fact, each digit automatically yields its predecessor and its successor. And the children have merely to go on tossing off (or reeling off) the digits one after another (forwards or backwards) by mere mental arithmetic (without needing pen or pencil, paper, slate, etc.).
5. On seeing this kind of work actually being performed by children, the doctors, professors and other 'big–guns' of Mathematics are wonder–struck and exclaim: Is this Mathematics or magic? And invariably the answer is: it is both. It is magic until we understand and it is Mathematics thereafter.
6. As regards the time required by the students for mastering the whole course of *Vedic* Mathematics as applied to all its branches, what is needed merely is, to state from the actual

experience, 8 months (or 12 months) at an average rate of 2 or 3 hours per day should suffice for completing the Mathematical studies on these *Vedic* lines.

7. Tough Mathematical problems (which the most advanced present day Mathematics need to spent huge amount of time, energy and money involving large numbers of difficult, tedious and cumbersome 'steps' of working) can easily and readily be solved with the help of these ultra–easy *Vedic Sūtra*–s in a few simple steps and methods. Hence this can conscientiously be described as "mental Mathematics".
8. It is thus in the fitness of things that the four *Vedas* – *Ṛg*, *Sāma*, *Yajus* and *Atharva* respectively include *upa–Vedas* as below, respectively:
   I. *Āyur Veda* (medical science).
   II. *Dhanur Veda* (archery and other military sciences).
   III. *Gāndharva Veda* (the science of art and music) and
   IV. *Sthapatya Veda* (engineering, architecture etc. and all branches of Mathematics in general).

All these subjects, it may be noted, are inherent parts of the *Vedas* i.e., are reckoned as 'spiritual' studies.

9. Similar is the case with *Vedāngas* (i.e., grammar, prosody, Astronomy, lexicography etc.) which according to the Indian cultural conceptions are also inherent parts and subjects of *Vedic* (i.e. religious) study.

4.2 Formulae and Corollaries of *Vedic* Mathematics

Vedic Mathematics has sixteen main formulae (*Sūtra*–s) and fourteen corollaries (*Upa–sūtra*–s or *Upa–Sūtra*–s). While there is no second opinion as to the number of formulae, the number and name of the corollaries differ in different schools. However, fourteen corollaries as detailed below have been accepted by majority of the schools. They are listed in the table below:

Table 5: *Vedic* Mathematics Formulae[4]

#	*Sūtra*	Formula in English	Translation
The main *Sūtra*–s – main formulae			
1.	एकाधिकेन पूर्वेण	*Ekādhikena Pūrveṇa*	By one more than the previous one
2.	निखिलं नवतश्चरमं दशत:	*Nikhilam Navathaścaramam Dhaśataḥ*	All from 9 and the last from 10.
3.	ऊर्ध्वतिर्यग्भ्याम्	*Ūrdhva Tiryagbhyām*	Vertically and Cross–wise
4.	परावर्त्य योजयेत्	*Parāvartya Yojayet*	Transpose and Apply
5.	शून्यं साम्यसमुच्चये	*Śūnyam Sāmyasamuccaye*	If the *Samuccaye* is the same it is Zero
6.	(आनुरूप्ये) शून्यमन्यत्	*(Ānurūpye) Śūnyamanyat*	If One is in Ratio the Other is Zero
7.	संकलनव्यवकलनाभ्याम्	*Sankalana–vyavakalanābhyām*	By Addition and by Subtraction
8.	पूरणापूरणाभ्याम्	*Pūraṇāpūraṇābhyām*	By the Completion or Non–Completion
9.	चलनकलनाभ्याम्	*Calana–kalanābhyām*	Differential Calculus

[4] This author has written a separate book titled "Vedic Mathematics", wherein all the 30 formulae have been explained individualy.

#	Sūtra	Formula in English	Translation
10.	यावदूनम्	*Yāvadūnam*	By the Deficiency
11.	व्यष्टिसमष्टि:	*Vyaṣṭisamaṣṭiḥ*	Specific and General
12.	शेषाण्यङ्केन चरमेण	*Śeṣāṇyaṅkena Carameṇa*	The Remainders by the Last digit
13.	सोपान्त्यद्वयमन्त्यम्	*Sopāntyadvayamantyam*	The Ultimate and Twice the Penultimate
14.	एकन्यूनेन पूर्वेण	*Ekanyūnena Pūrveṇa*	By One Less than the One Before
15.	गुणितसमुच्चय:	*Guṇitasamuccayaḥ*	The Product of the Sums
16.	गुणकसमुच्चय:	*Guṇakasamuccayaḥ*	All the Multipliers
	*The Sub or Upa–Sūtra–*s (corollaries)		
1.	आनुरूप्येण	*Ānurūpyeṇa*	Proportionately
2.	शिष्यते शेषसंज्ञ:	*Śiṣyate Śeṣasamjñaḥ*	The Remainder Remains Constant
3.	आद्यमाद्येन अन्त्यमन्त्येन	*Ādyamādyena Antyamantyena*	The First by the First and the Last by the Last
4.	केवलै: सप्तकं गुण्यात्	*Kevalaiḥ Saptakam Guṇyāt*	For 7 the Multiplicand is 143
5.	वेष्टनम्	*Veṣṭanam*	By Osculation
6.	यावदूनं तावदूनम्	*Yāvadūnam Tāvadūnam*	Lessen by the Deficiency

#	*Sūtra*	Formula in English	Translation
7.	यावदूनं तावदूनीकृत्य वर्गांच योजयेत्	*Yāvadūnam Tāvadūníkṛtya Vargañca Yojayet*	Whatever the Deficiency lessen by that amount and set up the Square of the Deficiency
8.	अन्त्ययोर्दशकेऽपि	*Antyayordaśake'pi*	Last Totaling 10
9.	अन्त्ययोरेव	*Antyayoreva*	Only the Last Terms
10.	समुच्चयगुणित:	*Samuccayaguṇitaḥ*	The Sum of the Products
11.	लोपनस्थापनाभ्याम्	*Lopanasthāpanābhyām*	By Alternative Elimination and Retention
12.	विलोकनम्	*Vilokanam*	By mere observation
13.	गुणितसमुच्चय: समुच्चयगुणित:	*Guṇitasamuccayaḥ Samuccayaguṇitaḥ*	The Product of the Sum is the Sum of the Products
14.	ध्वजाड	*Dvajāḍa*	On the flag

The *sūtra*–s embody laws, principles or methods of working and do not always easily succumb to rigid classification. Some of them have many applications. **Transpose and apply** is one such formula. It applies to solving equations, division in fractions and dividing numbers close to a base. It has many other uses at higher stages in Mathematics and indicates, not a single or particular algorithm, but a general mental procedure. There are other formulae, such as, **when the final digits add up to ten**, for which the uses appear to be very limited. It is because of the very few formulae which have many faceted qualities, that the subject becomes greatly unified and simplified.

More than one application may be used to solve a single Mathematical problem. Hence it is generally not possible to clearly list down which formula is to be used for a specific problem.

Normally corollaries are derived from the main theorems. But here, it can be observed that the *Upa–Sūtra*–s are not used under the main *sūtra*–s. They at times themselves act as individual *sūtra*–s. This is the main deviation from the usual Mathematical principles of theorem (*sūtra*) and corollaries (*Upa–Sūtra*).

The main reasons for this deviation could be:

- In general Mathematics – corollaries are for theorems and not for formulae.
- For want of better word we have translated the *upa–Sūtra*–s as corollary but in actual sense it is not a true meaning of that word and English is found wanting in this case for a better word to convey the exact meaning.

Actually a theorem or formula is not the exact English equivalent of the word '*sūtra*', for the word *sūtra* conveys more meanings like a laconic expression, when expanded yields many a meaning. But however for convenience *sūtra* continues to be called as formula.

*****

Annexure 1

Luminaries of Indian Mathematics

The history of Indian Mathematics dates back to the *Vedic* period (around 1500 B.C.E.). The *Śulba sūtra–s* of the *Vedic* age are texts on rules for altar construction. They are the oldest texts on Indian Mathematics. They contain the general enunciation of the Pythagoras theorem, approximate value for √2, methods of transforming one Mathematical figure to another, etc. Indians also used the decimal place–value system of representing the numbers. They had a representation for zero. The origin of the decimal place–value system in India was sometime around 1st Century B.C.E.

There are numerous great Mathematicians who have contributed to Indian Mathematics. The subject is a synergistic effort of all of them. Since even a mention of all of them would run into pages, the contributions of only a few of them are deliberated. The set below is only a drop in the ocean of great Mathematicians who lived in India. The exhaustive list itself is a separate candidate for research.

1. Jain Mathematicians

Although Jainism as a religion and philosophy predates its most famous exponent, Mahaveera (6th Century B.C.E.), who was a contemporary of Gautama Buddha, most Jaina texts on Mathematical topics were composed after the 6th Century B.C.E. Jaina Mathematicians are important historically as crucial links between the Mathematics of the Vedic period and that of the "Classical period".

A significant historical contribution of Jaina Mathematicians lay in their freeing Indian Mathematics from its religious and ritualistic constraints. In particular, their fascination with the enumeration of very large numbers and infinities, led them to classify numbers into three classes: enumerable, innumerable and infinite. Not content with a simple notion of infinity, they went on to define five different types of infinity: the infinite in one direction, the infinite in two directions, the infinite in area, the infinite everywhere and the infinite perpetually. In addition, Jaina Mathematicians devised notations for simple powers (and exponents) of numbers like squares and cubes, which enabled them to define simple algebraic equations (*beejganita samikaran*). Jaina Mathematicians were apparently also the first to use the word *shoonya* (literal meaning in Saṁskṛit is void) to refer to zero. More than a millennium later, their appellation became the English word "zero" after a tortuous journey of translations and transliterations from India to Europe.

In addition to *Surya Prajnapti*, important Jaina works on Mathematics included the *Vaishali Ganit* (3rd Century B.C.E.); the *Sthananga Sutra* (300 B.C.E. – 200 C.E.); the *Anoyogdwar Sutra* (200 B.C.E. – 100 C.E.); and the *Satkhandagama* (2nd Century C.E.). Important Jaina Mathematicians included Bhadrabahu (298 B.C.E.), the author of two astronomical works, the *Bhadrabahavi-Samhita* and a commentary on the *Surya Prajinapti*; Yativrisham Ācharya (176 B.C.E.), who authored a Mathematical text called *Tiloyapannati*; and Umasvati (150 B.C.E.), who, although better known for his influential writings on Jaina philosophy and metaphysics, composed a Mathematical work called *Tattwarthadhigama-Sutra Bhāshya*.

2. Pingala

Among other scholars of this period who contributed to Mathematics, the most notable is Pingala (*piṅgalá*) (300–200 B.C.E.), a musical theorist who authored the ***Chandaḥ-śāstra***, also

Chandaḥ-sūtra, a Saṁskṛit treatise on prosody. There is evidence that in his work on the enumeration of syllabic combinations, Pingala stumbled upon both the Pascal triangle and Binomial co-efficients, although he did not have knowledge of the Binomial theorem itself. Pingala's work also contains the basic ideas of Fibonacci numbers (called *mātrāmeru*). Although the *Chandah sutra* hasn't survived in its entirety, a 10th Century commentary on it by Halāyudha has. Halāyudha, who refers to the Pascal triangle as *Meru-prastāra* (literally "the staircase to Mount Meru"), has this to say:

"Draw a square. Beginning at half the square, draw two other similar squares below it; below these two, three other squares and so on. The marking should be started by putting **1** in the first square. Put **1** in each of the two squares of the second line. In the third line put **1** in the two squares at the ends and, in the middle square, the sum of the digits in the two squares lying above it. In the fourth line put **1** in the two squares at the ends. In the middle ones put the sum of the digits in the two squares above each. Proceed in this way. Of these lines, the second gives the combinations with one syllable, the third the combinations with two syllables".

The text also indicates that Pingala was aware of the combinatorial identity:

$$n_0 + n_1 + n_2 + n_3 + \text{-------} + n_{n-1} + n_n = 2^n$$

3. Āryabhaṭa

The world did not have to wait for the Europeans to awake from their long intellectual slumber to see the development of advanced Mathematical techniques. India achieved its own scientific renaissance of sorts during its classical era, beginning roughly 1000 years before the European Renaissance. Probably the most celebrated Indian Mathematician belonging to this period was Āryabhaṭa, who was born in 476 C.E.

In 499 C.E., when he was only 23 years old, Āryabhaṭa wrote *Āryabhaṭēya*, a text covering both Astronomy and Mathematics. With regard to the former, the text is notable for its awareness of the relativity of motion. This awareness led to the astonishing suggestion that it is the Earth that rotates around the Sun. He argued for the diurnal rotation of the earth, as an alternate theory to the rotation of the fixed stars and Sun around the earth. He made this suggestion approximately 1000 years before Copernicus.

His works inter–alia were in the areas of:

- Methods of determining square and cube roots
- Geometrical problems
- Solutions to problems relating to progressions
- Problems involving quadratic equations and indeterminate equations of the first degree. The method of solving these equations has been called *Kuṭṭaka* by later Mathematicians.
- He was the first Astronomer to mention that the apparent diurnal motion of the heavens is due to the rotation of the earth about its axis.
- Other contributions Āryabhaṭa made towards pure Mathematics were his sine tables, his approximation of π (*pi*) and the expressions that he gave for the sum of squares and the sum of cubes.

He calculated the value of π (*pi*) to four decimal places as 3.1416. In addition, Āryabhaṭa invented a unique method of recording numbers which required perfect understanding of zero and the place–value system.

The value of π (*pi*):

चतुरधिकं शतमष्टगुणं द्विषष्ठिस्तथा सहस्राणाम् ।
अयुतद्वयविष्कम्भस्य आसन्नो वृत्तपरीणाह ॥

Caturadhikam śatamaṣṭaguṇam dviṣaṣṭhistathā sahasrāṇām |
Ayutadvayaviṣkambhasya āsanno vṛtta parīṇāha || Āryabhaṭīyam

Add four to one hundred, multiply by eight and then add sixty two thousand then divide by twenty thousand – result is approximately circumference of a circle of diameter of twenty thousand.

Āryabhaṭa called it an approximate (*āsanna*) value.

π (*pi*) = circumference ÷ diameter = 62832 ÷ 20000 = 3.1416

Given the astounding range of advanced Mathematical concepts and techniques covered in 5th Century C.E., it should be of no surprise that it became extremely well known in India, judging by the large numbers of commentaries written upon it. It was studied by the Arabs in the 8th Century C.E., following their conquest of Sind and was translated into Arabic, whence it influenced the development of both Arabic and European Mathematical traditions.

With regard to Mathematics, one of the Āryabhaṭa's greatest contributions was the calculation of Sine tables, which no doubt was of great use for his Astronomical calculations. In developing a way to calculate the Sine curves, rather than the cruder method of calculating chords devised by the Greeks, he thus went beyond Geometry and contributed to the development of Trigonometry, a progress, which did not occur in Europe until roughly 1000 years later; when the Europeans translated Arab Mathematical texts sourced from India and translated into Arabic.

All his works are documented in *Āryabhaṭīya*.

4. Bhāskara

Bhāskara was one of the many outstanding Mathematicians hailing from South India. Born in 1114 C.E., in Karnataka, he composed a four–part text entitled the *Siddhānta Śiromaṇi*. Included in this compilation is the *Bīja Gaṇita*, which became the standard Algebra textbook in Saṁskṛit. It contains descriptions of advanced Mathematical techniques involving positive and negative integers as well as zero and irrational numbers. It treats at length the pulveriser (*kuṭṭaka*) method of solving indeterminate equations with continued fractions, as well as the Pell's equation (*varga prakṛti*) dealing with indeterminate equations of the second degree. He also gave solutions to numerous kinds of linear and quadratic equations, including those involving multiple unknowns and equations involving the products of different unknowns.

In short, he wrote a highly sophisticated Mathematical text that preceded several centuries the development of such techniques in Europe, although it would be better to term this a rediscovery, since much of the renaissance advances of Mathematics in Europe were based upon the discovery of Arab Mathematical texts, which were in turn highly influenced by these Indian traditions.

He was the head of the Ujjain observatory. His work was mainly in Algebra. He was influenced by Āryabhaṭa's work on *kuṭṭaka* and made his own modifications in order to solve indeterminate equations of the second degree. (Example: $Nx^2 + 1 = My^2$). This is known as the *cakravāla* process. Bhāskara wrote the *Siddhānta Śiromaṇi*, which has 4 parts to it –

- *Leelavati* (Arithmetic)
- *Beejagaṇita* (Algebra)
- *Grahagaṇita* (Mathematics of the planets) and
- *Golādhyāya* (celestial globe)

His famous works on Astronomy include *Siddhānta Śiromaṇi* and *Karaṇa Kuthūhalam*. The law of gravitation was propounded by him about 500 years before Newton.

In *Bījagaṇita*, a rough explanation of the concept of infinity occurs, in terms of the fact that any number divided by 0, gives infinity.

He had clear ideas of differential calculus. In the *Golādhyāya*, he gave the value of the ratio of the circumference of a circle to its diameter. The figure of an encryption shown below is claimed to be from his *Bījagaṇita*:

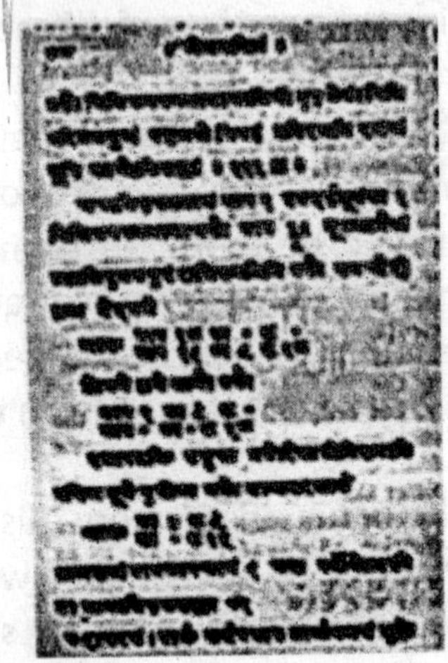

Figure 3: Encryption taken from *Bījagaṇita*

5. Kātyāyana

Though not a Jaina Mathematician, Kātyāyana (3rd Century B.C.E.) is notable for being the last of the *Vedic* Mathematicians. He wrote the *Katyayana Sulba Sutra*, which presented much geometry, including the general Pythagorean Theorem and a computation of the square root of 2 correct to five decimal places.

6. Brahmagupta

Born in 598 C.E., in Rajasthan in Western India, Brahmagupta founded an influential school of Mathematics which elaborated Āryabhaṭā's works. His best known work is the *Brahmaspuṭa Siddhānta*, written in 628 C.E., in which he developed a solution for a certain types of second order indeterminate equations. This text was translated into Arabic in the 8th Century C.E. and became very influential in Arab Mathematics.

He gave solutions to equations in Algebra like ($Nx^2 + y^2 +1 = 0$). These types of equations are known as "*Varga Prakṛti*". He also gave rules to solve simple quadratic equations and was the first to use Algebra in Astronomy. All his works are documented in the "*Brahmaspuṭa Siddhānta*'. It was through this work that the Arabs first heard of Astronomy. His contributions to Astronomy include calculations for the motions of various planets, the rising and setting of the Sun and the Moon and Solar and Lunar eclipses. Brahmagupta was one of the very few early Indian Mathematicians, who conceptualised the earth as round and not as flat and neither hollow.

He also gave numerous theorems for finding the area of a cyclic quadrilateral as:

A = square root of {(s – a) x (s – b) x (s – c) x (s – d)},
where a, b, c and d are the lengths of the sides and
s = ½(a + b + c + d).

7. Mahāvīrācārya

Mahāvīrācārya, a Jain Mathematician, lived in the 9th Century C.E., wrote on a wide range of Mathematical topics. These include the Mathematics of zero, squares, cubes, square–roots, cube–roots and the series extending beyond these. He also wrote on plane and solid Geometry, as well as problems relating to the casting of shadows.

The actual value of nCr in Permutation and Combination was given by Mahāvīrācāryā.

8. Mādhava

The Kerala region of South India was home to a very important school of Mathematics. The best known member of this school, Mādhava (1444 – 1545 C.E.) lived in Sangamagrāma in Kerala. Primarily an Astronomer, he made history in Mathematics with his writings on Trigonometry. He calculated the Sine, Cosine and Arctangent of a circle, developing the world's first consistent system of Trigonometry. He also correctly calculated the value of π (*pi*) to eleven decimal places.

9. Srīdharācārya

In the late 9th Century C.E., Srīdharācārya (probably Bengal) provided Mathematical formulae for a variety of practical problems involving ratios, barter, simple interest, mixtures, purchase and sale, rates of travel, wages and filling of cisterns. Some of these examples involved fairly complicated solutions and his *Pātigaṇita* is considered an advanced Mathematical work.

Sections of the book were also devoted to Arithmetic and Geometric progressions, including progressions with fractional numbers or terms and formulae for the sum of certain finite series. He gave solution for the equation $ax^2 + bx + c = 0$, which is being taught in schools even today.

10. Bhārati Krishna Teerthaji

Vedic Mathematics, compiled by, ***HH Jagadguru Swāmi Śrī Bhārati Kṛṣṇa Tīrthāji Mahāraj*** (जगद्गुरु स्वामि श्री भारती कृष्ण तीर्थाजी महराज: March, 1884 – February 2, 1960) of *Govardhan Pīṭha* is a monumental work. The phrase "**compiled by**" and not "written by" is consciously used, for the obvious reason that this was available in *Vedas* and *Śaṅkarācārya* discovered the formulae and did not invent them. In his deep–layer explorations of cryptic *Vedic* mysteries relating especially to the calculus of shorthand formulae and their neat and ready application to practical problems, the *Śaṅkarācārya* shows the rare combination of the probing insight of revealing intuition of a *Yogi* with the analytic acumen and synthetic talent of a Mathematician.

P.S. For continuity sake this paragraph has been mentioned here though detailed chapter on Vedic Mathematics is given separately.

*****

Annexure 2

Table 6: List of *Melakarta Rāgās*

#	*Rāga* in English	*Rāga* in Saṁskṛit
1	*Kanakāngi*	कनकांगि
2	*Ratnāngi*	रत्नांगि
3	*Gānamūrti*	गानमूर्ति
4	*Vanaspathi*	वनस्पथि
5	*Mānavathi*	मानवथि
6	*Thānarūpi*	थानरूपि
7	*Senāvathi*	सेनावथि
8	*Hanumathoḍhi*	हनुमथोढि
9	*Dhenukā*	धेनुका
10	*Nāṭakapriyā*	नाटकप्रिया
11	*Kokilapriyā*	कोकिलप्रिया
12	*Rūpavathi*	रूपवथि
13	*Gāyakapriyā*	गायकप्रिया
14	*Vakulābharaṇam*	वकुलाभरणम्
15	*Māyāmālavagoula*	मायामालवगौल
16	*Cakravākam* **	चक्रवाकम्
17	*Sūryakāntam*	सूर्यकान्तम्
18	*Haṭakāmbari*	हटकाम्बरि
19	*Jhankāradhvani*	झन्कारध्वनि
20	*Naṭabhairavi*	नटभैरवि
21	*Kīravāṇi*	कीरवाणि
22	*Kharaharapriyā*	खरहरप्रिया
23	*Gourimanohari*	गौरिमनोहरि
24	*Varuṇapriyā*	वरुणप्रिया
25	*Māraranjani*	माररन्जनि

#	*Rāga* in English	*Rāga* in Saṁskṛit
26	*Cārukeśi*	चारुकेशि
27	*Sarasāngi*	सरसान्गि
28	*Harikāmbhoji*	हरिकाम्भोजि
29	*Dhīraśankarābharaṇam*	धीरशन्कराभरणम्
30	*Nāganandini*	नागनन्दिनि
31	*Yāgapriyā*	यागप्रिया
32	*Rāgavardhani*	रागवर्धनि
33	*Gāngeyabhūśani*	गान्गेयभूशनि
34	*Vāgadhīsvari*	वागधीस्वरि
35	*Śūlini*	शूलिनि
36	*Calanāta*	चलनात
37	*Sālagam*	सालगम्
38	*Jalārnavam*	जलार्नवम्
39	*Jhālavarāḷi*	झालवराळि
40	*Navanītham*	नवनीथम्
41	*Pāvani*	पावनि
42	*Raghupriyā*	रघुप्रिया
43	*Gavāmbodhi*	गवाम्बोधि
44	*Bhavapriyā*	भवप्रिया
45	*Śubhapanthuvarāḷi*	शुभपन्थुवराळि
46	*Śadhvidha Mārgiṇi*	शध्विध मार्गिणि
47	*Suvarṇāngi*	सुवर्णांगि
48	*Dhivyāmaṇi*	धिव्यामणि
49	*Dhavaḷāmbari*	धवळाम्बरि
50	*Nāmanārāyaṇi*	नामनारायणि
51	*Kāmavardhini*	कामवर्धिनि
52	*Rāmapriyā*	रामप्रिया

#	*Rāga* in English	*Rāga* in Saṁskṛit
53	*Gamanaśrama*	गमनश्रम
54	*Viśvāmbhari* **	विश्वाम्भरि
55	*Śyāmaḷāngi* **	श्यामळान्गि
56	*Ṣaṇmukhapriyā*	षण्मुखप्रिया
57	*Simmendra Madhyamam***	सिम्मेन्द्र मध्यमम्
58	*Hemāvathi*	हेमावथि
59	*Dharmāvathi*	धर्मावथि
60	*Nīthimathi*	नीथिमथि
61	*Kāntāmaṇi*	कान्तामणि
62	*Riśabhapriyā*	रिशभप्रिया
63	*Lathāngi*	लथान्गि
64	*Vācaspathi*	वाचस्पथि
65	*Mecakalyāṇi*	मेचकल्याणि
66	*Citrāmbari* **	चित्राम्बरि
67	*Sucarithā*	सुचरिथा
68	*Jyotisvarūpiṇi* **	ज्योतिस्वरूपिणि
69	*Dātuvardhani*	दातुवर्धनि
70	*Nāsikā Bhūśaṇi*	नासिका भूशणि
71	*Kosalam*	कोसलम्
72	*Rasikapriyā*	रसिकप्रिया

** Earlier it was stated that the half letters are not accounted for in all Mathematical calculations. The ragas marked ** above are some exceptions to this rule. The half letters are also considered in these cases.

*****

Annexure 3

The numerals corresponding to each of the Saṁskṛit alphabets used in *Āryabhaṭīya Saṅkhyā*:

Table 7: The Numerals Relating to Saṁskṛit Alphabets

The $33 \times 9 = 297$ Saṁskṛit alphabetic numerical syllables											
Nine vowels or syllabics			*a*	*i*	*u*	*r*	*l*	*E*	*ai*	*o*	*Au*
			अ	इ	उ	ऋ	ऌ	ए	ऐ	ओ	औ
×			10^0	10^1	10^2	10^3	10^4	10^5	10^6	10^7	10^8
			Eka	*Dasa*	*Śata*	*Sahasra*	*Ayuta*	*Lakṣa*	*Prayuta*	*Koti*	*Nyadbhuta*
Five velar plosives											
k	क	1	क *ka*	कि *ki*	कु *ku*	कृ *kṛ*	कॢ *kḷ*	के *ke*	कै *kai*	को *ko*	कौ *Kau*
Kh	ख	2	ख *kha*	खि *khi*	खु *khu*	खृ *khṛ*	खॢ *khḷ*	खे *khe*	खै *khai*	खो *kho*	खौ *khau*
g	ग	3	ग *ga*	गि *gi*	गु *gu*	गृ *gṛ*	गॢ *gḷ*	गे *ge*	गै *gai*	गो *go*	गौ *gau*

gh	घ	4	घ gha	घि ghi	घु ghu	घृ ghṛ	घॢ ghḷ	घे ghe	घै ghai	घो gho	घौ ghau
ṇ	ङ	5	ङ ṇa	ङि ṇi	ङु ṇu	ङृ ṇṛ	ङॢ ṇḷ	ङे ṇe	ङै ṇai	ङो ṇo	ङौ ṇau
Five palatal plosives											
c	च	6	च ca	चि ci	चु cu	चृ cṛ	चॢ cḷ	चे ce	चै cai	चो co	चौ cau
ch	छ	7	छ cha	छि chi	छु chu	छृ chṛ	छॢ chḷ	छे che	छै chai	छो cho	छौ chau
j	ज	8	ज ja	जि ji	जु ju	जृ jṛ	जॢ jḷ	जे je	जै jai	जो jo	जौ jau
jh	झ	9	झ jha	झि jhi	झु jhu	झृ jhṛ	झॢ jhḷ	झे jhe	झै jhai	झो jho	झौ jhau
ṅ	ञ	10	ञ ṅa	ञि ṅi	ञु ñu	ञृ ṅṛ	ञॢ ñḷ	ञे ṅe	ञै ṅai	ञो ṅo	ञौ ṅau
Five retroflex plosives											
ṭ	ट	11	ट ṭa	टि ṭi	टु ṭu	टृ ṭṛ	टॢ ṭḷ	टे ṭe	टै ṭai	टो ṭo	टौ ṭau
ṭh	ठ	12	ठ ṭha	ठि ṭhi	ठु ṭhu	ठृ ṭhṛ	ठॢ ṭhḷ	ठे ṭhe	ठै ṭhai	ठो ṭho	ठौ ṭhau

ḍ	ड	13	ड ḍa	डि ḍi	डु ḍu	डृ ḍṛ	डॢ ḍḷ	डे ḍe	डै ḍai	डो ḍo	डौ ḍau
ḍh	ढ	14	ढ ḍha	ढि ḍhi	ढु ḍhu	ढृ ḍhṛ	ढॢ ḍhḷ	ढे ḍhe	ढै ḍhai	ढो ḍho	ढौ ḍhau
ṇ	ण	15	ण ṇa	णि ṇi	णु ṇu	णृ ṇṛ	णॢ ṇḷ	णे ṇe	णै ṇai	णो ṇo	णौ ṇau
Five dental plosives											
t	त	16	त ta	ति ti	तु tu	तृ tṛ	तॢ tḷ	ते te	तै tai	तो to	तौ tau
th	थ	17	थ tha	थि thi	थु thu	थृ thṛ	थॢ thḷ	थे the	थै thai	थो tho	थौ thau
d	द	18	द da	दि di	दु du	दृ dṛ	दॢ dḷ	दे de	दै dai	दो do	दौ dau
dh	ध	19	ध dha	धि dhi	धु dhu	धृ dhṛ	धॢ dhḷ	धे dhe	धै dhai	धो dho	धौ dhau
n	न	20	न na	नि ni	नु nu	नृ nṛ	नॢ nḷ	ने ne	नै nai	नो no	नौ nau

Five labial plosives											
p	प	21	प *pa*	पि *pi*	पु *pu*	पृ *pṛ*	पॢ *pḷ*	पे *pe*	पै *pai*	पो *po*	पौ *pau*
ph	फ	22	फि *pha*	फि *phi*	फु *phu*	फृ *phṛ*	फॢ *phḷ*	फे *phe*	फै *phai*	फो *pho*	फौ *phau*
b	ब	23	ब *ba*	बि *bi*	बु *bu*	बृ *bṛ*	बॢ *bḷ*	बे *be*	बै *bai*	बो *bo*	बौ *bau*
bh	भ	24	भ *bha*	भि *bhi*	भु *bhu*	भृ *bhṛ*	भॢ *bhḷ*	भे *bhe*	भै *bhai*	भो *bho*	भौ *bhau*
m	म	25	म *ma*	मि *mi*	मु *mu*	मृ *mṛ*	मॢ *mḷ*	मे *me*	मै *mai*	मो *mo*	मौ *mau*
Four approximants or trill											
y	य	30	य *ya*	यि *yi*	यु *yu*	यृ *yṛ*	यॢ *yḷ*	ये *ye*	यै *yai*	यो *yo*	यौ *yau*
r	र	40	र *ra*	रि *ri*	रु *ru*	रृ *rṛ*	रॢ *rḷ*	रे *re*	रै *rai*	रो *ro*	रौ *rau*
l	ल	50	ल *la*	लि *li*	लु *lu*	लृ *lṛ*	लॢ *lḷ*	ले *le*	लै *lai*	लो *lo*	लौ *lau*
v	व	60	व *va*	वि *vi*	वु *vu*	वृ *vṛ*	वॢ *vḷ*	वे *ve*	वै *vai*	वो *vo*	वौ *vau*

Three coronal fricatives											
ś	श	70	श *śa*	शि *śi*	शु *śu*	शृ *śṛ*	शॢ *śḷ*	शे *śe*	शै *śai*	शो *śo*	शौ *śau*
ṣ	ष	80	ष *ṣa*	षि *ṣi*	षु *ṣu*	षृ *ṣṛ*	षॢ *ṣḷ*	षे *ṣe*	षै *ṣai*	षो *ṣo*	षौ *ṣau*
s	स	90	स *sa*	सि *si*	सु *su*	सृ *sṛ*	सॢ *sḷ*	से *se*	सै *sai*	सो *so*	सौ *sau*
One glottal fricative											
h	ह	10 0	ह *ha*	हि *hi*	हु *hu*	हृ *hṛ*	हॢ *hḷ*	हे *he*	है *hai*	हो *ho*	हौ *Hau*

*****

Bibiliography

Referenced texts/ authors:

#	Name	Author	Publications
1.	A History of the Kerala School of Hindu Astronomy	Sarma K V	Hoshiarpur, 1972
2.	A Universal History of Numbers: From Prehistory To The Invention Of The Computer	Ifrah G	London, 1998
3.	An Overview Of Ancient Indian Science	Subhash Kak	Munshiram Manoharlal Publishers Pvt. Ltd. 2001
4.	Contributions Of The Jainas To Astronomy And Mathematics	Jha P	Math. Ed. (Siwan) 18 (3) (1984)
5.	From One To Zero	Ifrah, Georges	New York: Viking Penguin
6.	Geometry In Ancient And Medieval India	Saraswati T.A.	1979, Delhi
7.	Hindu Mathematics In The 7th Century C.E., As Found In Bhāskara I's Commentary On The Āryabhaṭēya,	Shukla K.S.	Gaṇita Bharati

IMPORTANT BOOKS ON INDIAN ASTROLOGY

ENCYCLOPAEDIA OF INDIAN ASTROLOGY

PROF.N.E.MUTHUSWAMY. *M.A.M.Ed, Nyamamahopadhyaya, Sahithyaratna, D.I.M*

A comprehensive Encyclopaedia on Indian Astrology comprising a detailed description of the concepts encom-passing the various branches of Indian Astrology, Predictions, Combinations, Prasna, Ashtamangalaprasna, Stellar Astrology, Muhoortha, etc.

Highlights of the encyclopaedia

More than 300 articles,15,000 words relating to Indian Astrology, 300 tables of various astrological data, comprehensive and exhaustive information on Prasna and Ashtamangala branch of Indian Astrology, different systems of Indian Astrology, Combinations based on various astrological concepts extracted from Indian classics on Astrology in Sanskrit like Parasara hoara, Jathakapaarijatja, Brihathjathakam, Varahahoara, Saaravali, Krishneeyam, Prasnamarga and Tamil classics on Indian astrology.

The well edited articles on various branches of Indian Astrology, various planetary combinations and possible predictions based on these planetary positions and other aspects are dealt exhaustively in the encyclopaedia.The various astrological combinations and their results enable the easy prediction of astrological results. More than10000 combinations analysed with prediction.

"Most Exhaustive reference book on Indian Astrology released to date." Absolutely essential reference book for every Indian astrologer, learner and the practioner".

Crown quarto / 2 volumes / 2006/1200 pages / Price for set of 2 volumes Rs.1800.00

Prof.N.E.Muthuswamy, is known in astrological circles in Kerala as the Upakulapathi of Indian Astrology. He had been ardently promoting the learning of Indian astrology among the various sections of Kerala for the last 15 years and he had a large student base. He has written more than 60 books on astrology in Malayalam and has written Exhaustive commentary for Saravali, Brihathparasarahoar, Uthara Kaalamritham, Prasnanushtanapadhathi, Prasnamarga, and critically edited with text and commentaries on several books on various aspects of Indian astrology. His books in English, Course in Indian Astrology, Gulika in Astrology, Ashtamangala Prasna, Medical Astrology have been well received.

1350 PAGES/CROWN 1/4TH SIZE / IN 2 VOLUMES / Rs 2000 for set of 2 volumes

COURSE IN INDIAN ASTROLOGY

In the present book the materials have been arranged in simple style comprising of smaller units in a systematic and comprehensive manner to enable the learner to learn the South Indian system of astrology.Well arranged tables and exercises takes the learner through to the various concepts of Indian System of Astrology. North Indian and South Indian system of charts have been incorporated in the text.

"A Complete reader in itself " *Rs 150.00*

MEDICAL ASTROLOGY

vol 1 Concepts and case studies, vol 2 Remedial measures

A complete reader on Medical astrology. The book deals with the application of Indian astrology to Medical aspects and treatment of diseases and relates them to principles of Indian astrology. More than 100 case studies are analysed in detail. A complete text book on Medical Astrology. Contains extensive chapters on remdial measures as praticed in astrology as well.

First volume, Concepts and case studies -in Medical Astrology -- Starting with the Basic concepts of Indian astrology as relevant to Medical Astrology are presented. Then a set of case studies are presented for understanding the application of astrological principles and how they could be used for analyzing the diseases and their onset for a native, as available from horoscopes.

Second volume - Remedial measures in Medical astrology starts with different types of diseases and how the astrological remedies are available for these diseases. The planetary combinations for different diseases are well documented. The book goes on to present astrological remedies as are available in Prasna Marga, Karma Vipakam, Lal Kitab, Veerasimhavalokam etc. A Detailed remedies for different diseases as available through Navagraha dosha shanthi, Gemmology etc are well documented. *Vol 1 Rs 250.00 / 225 pages, Vol 2 Rs 250.00 /256 pages*

ASHTAMANGALAPRASNA

An important Branch of Indian Astrology. A complete detailed study of the Ashtamangalaprasna has been undertaken in this volume. The book goes in detail about the intricacies of conducting an Ashtamangalaprasna, selecting and analysing the aroodta, its revelation of various details during the prasna, solution to various day to day problems, diseases,etc. *Rs 150.00*

GULIKA IN ASTROLOGY

Gulika is a sensitive point in the Zodiac emerging at different points on different days. The rich and varied effects, good and bad in one's life are attributed to Gulika in Astrology. The present book makes an in-depth of study of Gulika and its various aspects based on the classics of Indian and Kerala Astrology, like Prasnamarga, Phaladeepika etc. Extensively quoted with Sanskrit texts in Devanagari. This is a compendium of all relevant material available on Gulika. *Rs.75.00*

"The only study of its kind"

BOOKS BY V.S.KALYANRAMAN

INDIAN ASTROLOGY - AN APPRAISAL-

The book is an inspiring compilation of interesting and informative articles on the various characteristics of Indian Astrology.Surveying in brief the land's contribution to the astrological erudition, it delineates on some of the eminent scholars it had produced in the past. Discussing the causes, the book can serve as a perennial guide to any practitioner besides enlightening the student and engaging the scholar. *Rs 170.00*

YOGARNAVA

This book is a compilation of YOGAS that are specific planetary combinations, identifiable in any horoscope, which are deemed to have the special power and capacity of bringing forth some explicit results, good or bad. They are found in abundance and lie scattered by mention, in the various astrological texts of yore. Through an incisive evaluation, the book helps the reader to appraise the yogas in astrology appropriately. In addition it engulfs itself with an encyclopaedic content of many informative inputs. In short it is a vade-mecum for anyone, whose passion is astrology. *Rs 325.00*

MUHOORTHA SANGRAHA

This book in 2 volumes comes in very handy to a layman to ubnderstand how a suitable time for each activity is selected. The present effort is to study in detail and collate the volley of information found scatted in different texts on the Muhoortha branch of astrology. The book analyses the significant minutae of muhoortha and present them cogently in the form a handbook. The first volume gives in details various principles and theory of finding a suitable muhoortha, as explained in various calssical astrology tests. The second volume presents the various pitfal in arriving the muhoortha, and a complete details on various muhoortha for different occassions. The book elucidates the fundamental principles of arriving at an acceptable muhoortha for every concievable occassion in the present day. Vol 1 Rs 250.00 / 225 pages, Vol 2 Rs 250.00 /256 pages

DASA SYSTEM AND PREDICTION

In this book, a volley of major ancient texts like, *Paraasara Samhita*, *Sarvaartha Chintamani*, *Phaladeepika* and the Tamil classic *Jathaka Alankaaram*, have been mainly depended upon to enunciate the results. Introducing the basic concepts of the Dasa system, a deep discussion on Udu dasa system, which is the present prevalent system on which the dasa prediction is based and is in practice. The introduction deals with an overview of the different dasa systems in Indian astrology as culled out from different classic astrology texts. The second section goes into general dasa prediction for each Planet and its positioning in each rasi and then delves with direct prediction for sub period and antardasas of each planet.The predictions are so arranged that the book can be used as ready reference to arrive at the results for each planet and its sub periods. The book is intended to be a ready-reckoner, to the students and practitioners alike. /2014 Rs 400.00

STHREEJATHAKA

DR.R.NATARAJAN

Core subject of this Book is matching horoscopes for a happy married life. This is done by Comparing Doshas and matching the Nakshatras. In this book the Author has discussed the Stree

Jataka in respect of their longevity, health, wealth, education, profession, appearance, physique, character,family, relationships, friends etc. Similar aspects of the husband also have been discussed with reference to the Kalathra Sthana (or 7th House) of the female horoscope. The type of marriage the native will have whether it will be love-marriage, registered marriage, inter-cast or inter- religion marriage or conventional arranged marriage have also been discussed. Matching is discussed in Chapter 9, based on ten types of agreements corresponding to the Nakshatras of the female and the male. The importance of each of the agreement also has been brought out which, astrologers normally do not mention. The marriage is not only to get life-partner but also to have good children. In Chapter 10, the method of choosing auspicious time is discussed for the union of husband and wife so that good children are born. The last chapter details the progeny of the family, with analysis of different horoscopes in particular. To facilitate this matching a Ready Reckoned Table is given in the Appendix. 2014/ Rs 400.00 /

LONGEVITY

P.M.GOPALACHARY

This book on Longevity is an extended study on the principles of Longevity in Indian Astrology. Starting with the various concepts as depicted in different texts, the author analyses the different types, combinations, nature of different ayurdayas. The different causes for longevity of the native like place, nature, etc. are correlated with live examples of personalities. The Basic tenets for determining the longevity from drekkanas, Jaimini astrology, pidayu method, kalachakra dasa, ashtakavarga, Krishnaorrthy padhathi, are enunciated. The influence of Nakshatras, maraka periods in the natives life, are explained with examples of horoscopes. the importance of the seventh house in one's longevity is well brought out. This book will guide the learner as well as the practicing astrologer to arrive at a plausible life term of a native by proper use of this book.

PLANETARY DISPOSITIONS AND MARITAL AFFAIRS

Dr.V.P.Shukla

The first book of its kind based on research oriented papers. Marriage is a landmark in one's life. A blissful marital life is an asset while a disturbed and ruined conjugal life is only a bed of thorns. In order to help ensure a happy marital life, an effort has been made to analyse planetary influences on marital affiars.The theme of the book is to educate our readers and warn them of certain factors governing the fragility of marriages which could be averted or managed properly. It is hoped the book could help the readers and astrologers as well. *Rs 210.00*

KALAPRAKASIKA

N.P.SUBRAMONIA IYER

The important classic on Muhortha, the electional astrology. The book gives a comprehensive introduction to the calculation of muhoortha for various electional muhurthas. A copy of the ephemeris and the kalaprakasika is the only thing an astrologer would need to give best muhoorthas for every known functions and astroligical rituals.An essential book for every astrologer Rs 175.00

VIDEO LECTURE SERIES ON INDIAN ASTROLOGY

A SERIES OF WELL DOCUMENTED VIDEO LECTURE SERIES IN EASY TO LEARN MODULES FOR EASY AND FASTER LEARNING THE CONCEPTS OF ASTROLOGY

PRESENTED BY SENIOR ACADEMICS AND PRACTICING ASTROLOGERS IN DVD FORMAT WITH POWER POINT PRESENTATIONS AND VISUALS

Lessons are presented by Prof.R.M.Palaniappan and V.S.Kalyanraman

1.BASIC CONCEPTS - by Prof.R.M.Palaniappan

The Basic concepts of Indian Astrology namely the thithi, vara, nakshatra concepts, the panchang, the zodiac, and its importance, the rasi concepts, the hoses and the bhavas, the horoscope, different houses in a horoscope, the fundamental concepts of casting the horoscope etc. Are well explained with power point Presentations and descriptive lectures.

2.ANALYSIS OF HOUSES - by V.S.Kalyanraman

The lecture on analysis of houses, presents a basic Introduction to the houses in a horoscope and evaluates the qualities of each house. The principles are evaluated on the basis of a critical analysis with respect to different charts and how a horoscope is to be analysed depending on the planetary positions.

3.THE ASCENDANT AND THE HOUSES - by Prof.R.M.Palaniappan

The Ascendant forms the basis for delineation of houses in Indian astrology. This important facet is the basis on which the predictive principles of astrology revolves about. In this video lecture, the concepts of Ascendant, various houses, results of planetary placement for different ascendants, are well presented for the beginner.